PRACTICAL PETROLEUM GEOLOGY

by
Jeff Morris
Richard House
Annes McCann-Baker

Edited by
Jodie Leecraft

Published by

Petroleum Extension Service
Division of Continuing Education
University of Texas at Austin
Austin, Texas

in cooperation with
Association of Desk and Derrick Clubs
Tulsa, Oklahoma
1985

The University of Texas at Austin is an equal opportunity institution. No state tax funds were used to print or mail this publication.

Catalog No. 1.00210
ISBN 0-88698-097-6

The University of Texas at Austin is an equal opportunity employer. No state funds were used to produce this manual.

Contents

FOREWORD

The Association of Desk and Derrick Clubs is an educational organization comprised of 10,000 members employed in the petroleum and allied industries. These members are convinced that by acquiring greater knowledge of their multifaceted industry, they can enlarge their interests and thereby improve their scope of service.

To further their goals, Desk and Derrick developed, in conjunction with the Petroleum Extension Service of The University of Texas at Austin (PETEX), a popular textbook, *Fundamentals of Petroleum,* an overview of the basic aspects of the petroleum industry, which is now in its third printing.

To expand this line of publications to cover each strategic portion of the industry in more depth, a second volume encompassing land and leasing was compiled from the ideas and guidelines presented to PETEX by Desk and Derrick. This second volume, *Land and Leasing,* has received wide acceptance from the membership.

The third volume from this joint effort of ADDC and PETEX is *Practical Petroleum Geology,* designed to explain the origin and related characteristics of petroleum sources to the non-professional in broader scope and greater detail than *Fundamentals of Petroleum.*

The Association of Desk and Derrick Clubs has very positive concepts on the value of continuing education within the petroleum and allied industries and is pleased to offer these books to its members and the businesses they represent.

Association of Desk and Derrick Clubs
315 Silvey Building
Tulsa, Oklahoma 74119

PREFACE

In developing plans for the writing of *Practical Petroleum Geology,* the Association of Desk and Derrick Clubs and Petroleum Extension Service decided that the book, whose purpose was to present a description of the field of petroleum geology, would be centered on the point of view of the practicing petroleum geologist. From this decision came the word *practical* in its title—not practical in the sense of conveying how-to directions to the reader, but practical in the sense of being down-to-earth and based on the understanding, skill, and knowledge actually required of a petroleum geologist in pursuing his daily work.

Presumably such a book did not exist. And to Desk and Derrick members who worked in organizations involved primarily or peripherally with petroleum geology, such a book was visualized as something very useful. PETEX perceived its benefits as extending to an even wider circle of readers, including those working in other areas of the petroleum industry and those wishing to prepare themselves for some geology-oriented career and eager to learn about this particular one.

Organization of the book seemed to grow naturally from its content and point of view. It begins at the beginning, with the basic concepts that any student of geology must know. From among those concepts, sedimentation, the geological process that is the focus of interest for the petroleum geologist, is singled out for more detailed discussion. And then the processes that result in the formation, accumulation, and migration of oil and gas are explained.

After the beginning three chapters of pure geology, slanted of course toward the petroleum geologist's interest in oil and gas, the discussion turns toward more active concerns. The petroleum geologist is vitally concerned in the activity of exploration; in fact, he is essential to this endeavor. The fourth chapter covers his primary tasks here: locating a prospect, making geological surveys, and documenting the prospect.

After exploration has brought the geologist to the point of recommending whether to drill a prospect or not, he must, with the managers of his company or with possible investors in the drilling project, take time for a realistic appraisal of economic pros and cons. The concepts needed to understand how such decisions can be rationally and fruitfully made are explained in the fifth chapter. Although the geologist does not make such decisions alone, he is usually an active member of the group that does.

The exploratory well follows an affirmative decision to drill. Here the petroleum geologist assumes a major role, that of well sitting. His responsibilities during this time, including lithological logging of the well, are described in chapter 6.

Production from the exploratory well initiates field development, a task that is broad and complex, requiring a great deal of input from the geologist's knowledge of the reservoir. It is with this study in chapter 7 that the book ends its look at the petroleum geologist's field of expertise and work.

Nothing can be seen whole, however, without a context. And so the final chapter offers a perspective—a look at petroleum geology in the contexts of history, the petroleum industry, and the use of the resource for which it exists.

Practical Petroleum Geology was written by three staff writers of PETEX: Jeff Morris, responsible for the first four chapters; Richard House, for the last three; and Annes McCann-Baker, for chapter 5. They worked diligently on both the research and the writing, sought eagerly for frequent reviews of their work by competent professionals in the field, and checked their data carefully. But errors and omissions almost surely remain. The authors and editor would welcome having those brought to their attention.

The book was an enlightening and interesting project to prepare for publication. We hope that it proves equally enlightening and interesting to the reader.

Jodie Leecraft
Editor

ACKNOWLEDGMENTS

We at Petroleum Extension Service are truly grateful to those people who helped so considerably in reviewing our manuscript as it developed, guiding our writers with suggestions based on their significant professional experience and knowledge. They were generous in their gifts of time and effort, and we offer our heartfelt thanks to the following:

Walter Berryhill, Chevron Geosciences, Houston, Texas

J. D. Broomell, ARCO Oil and Gas Company, Dallas, Texas

Samuel P. Ellison, Jr., Professor of Geological Sciences, The University of Texas at Austin

Redge Greenberg, Oiltex International, Austin, Texas

E. F. Herbeck, ARCO Oil and Gas Company, Dallas, Texas

R. E. Megill, Kingwood, Texas

W. F. Smylie, BASA, Inc., Kilgore, Texas

We also wish to thank Thomas Liston of Texas International Exploration Company of Houston, who, along with Walter Berryhill, spent several fruitful sessions of discussion with us in planning what the book should cover and how it should be organized. Our former colleague Mark Longley was also very much involved in the planning stage, and we acknowledge his helpful ideas.

The Railroad Commission of Texas was most cooperative in clarifying several points and furnishing their forms, and the Bureau of Economic Geology of The University of Texas at Austin was likewise helpful in clarifying other points and furnishing several excellent illustrations.

Finally, we should like to give special thanks to Lewis Raymer of Schlumberger Well Services in Houston for his inestimable contribution in drawing a well log response chart for chapter 6. He was most gracious in answering our urgent need.

Artwork for the book was done by J. Kay Wilson; layout, by Debbie Caples; and typesetting, by Cindy Carrell.

Figure 1.1. Eruption of Mount St. Helens, May 18, 1980 (Courtesy of United States Geological Survey)

BASIC CONCEPTS OF GEOLOGY

What comes into your mind when you hear the word *geology*? If you're like most of us, chances are you get a vivid picture—such as the May 1980 eruption of Mount St. Helens (fig. 1.1). We tend to think of geology in terms of landscapes too vast to fully comprehend—volcanoes, mountain ranges, canyons—created by forces beyond our control.

Perhaps, then, you will be surprised to learn that most geological changes occur so slowly that you cannot see them happening; that our understanding of those changes has enabled us to find the fossil fuels that power our civilization; and that the scope of our insight into the geological phenomenon of oil includes both the large and the small— mountains and sand grains, oceans and water droplets, sunlight and bacteria.

Within this wide range of things and ideas, the story of how, where, and when petroleum accumulates underground can be made very complicated—or very simple. A scientist can describe precisely how heat and pressure affect complex organic molecules, but he may do so in a language hard to understand. The average motorist, on the other hand, may fill up at the gas pump, comfortable with the notion that oil comes from dinosaur-shaped caverns.

There is a grain of truth (and a mountain of myth) in the latter account; an abundance of precision (and obscurity) in the former.

For those curious about the phenomenon of nature's liquid fuels but lacking a rigorous scientific education, the simple picture does not satisfy, but the exact answer (insofar as scientists know it) baffles.

Suppose you were to ask the nearest person why it rains. A first grader might say, "The clouds break open." A high school student might explain that water evaporates from the oceans, is carried inland by winds, and cools and condenses over the land. A college graduate might expound at length on prevailing westerlies and local convection cells and adiabatic cooling and condensation of supercooled water molecules on microscopic dust particles. A nuclear physicist, overhearing all this, might feel constrained to tell you that the adsorption of water molecules on dust particles is a function of the exchange of pi-mesons among protons—which is really more than you wanted to know.

You can see right away, however, that a simple, yet accurate, explanation of a natural phenomenon such as rainfall requires you to examine events at many levels—microscopic, local, planetary. Each level is important. Showing how water vapor condenses on dust particles does not explain how it got up into the cloud tops; describing how water evaporates from the oceans fails to account for the barrenness of the Sahara.

Explaining the occurrence of petroleum in rock formations is much the same, with one fundamental difference—we can watch clouds forming and rain falling, but no one has ever watched oil form and accumulate underground (fig. 1.2). To figure out what happens, we must first answer some basic questions: Does oil start out as oil or as something else? Does it change over time? Does it form where it is found, or does it come from somewhere else? Why is it found in some places but not in others? These questions can be answered; but to answer them we have to look at petroleum on many levels—global, regional, local, microscopic, and molecular.

Figure 1.2. Oil and rain

Scientists collect facts and construct theories to explain them. Cause and effect, action and consequence are examined in minute detail and in universal scope. These observations and theories are grouped into relevant areas. In the search for petroleum, the area of greatest interest to us is geology—the science of the earth.

UNIFORMITARIANISM

Perhaps the easiest way to understand how and where petroleum originates and accumulates is to study processes. At every level, nature is constantly changing. Some of these changes may alter the landscape suddenly and spectacularly, as eyewitnesses to the 1980 explosion of Mount St. Helens can attest—those who survived. Most changes, however, occur so gradually that we are not aware they are happening. A visitor to the Grand Canyon will see, as far as he can tell, the same canyon in 1985 that he saw in 1935. Yet gradual change can alter our world as profoundly as natural cataclysms—and the formation and accumulation of petroleum is a gradual process.

Geologic processes can be used not only to explain how the earth looks today, but also to deduce how it looked millions of years ago. To do so, we must make one broad assumption—that the processes that are at work today on the earth are the same as, or very similar to, the processes that affected the earth in the past. This principle, known as *uniformitarianism* (or *gradualism*), was formulated in 1785 by James Hutton, a Scottish geologist. It became an accepted doctrine with the publication in 1830 of English geologist Charles Lyell's book *Principles of Geology*. Another way of stating the principle is "the present is the key to the past."

Before Hutton and Lyell, the prevailing view (based more on religious dogma than field evidence) was that the earth's landforms had assumed their present configuration in a brief episode at the beginning of geologic history, possibly in a single great catastrophic event, and had remained relatively unchanged since that time. This theory, sometimes called *catastrophism*, failed to account for such familiar features as thick layers of sedimentary rock that geologists now know could only have accumulated during millions of years of deposition in a variety of geographic settings.

The landforms of the earth today are the result of an accumulation of similarly small, often imperceptible changes occurring over a long period according to essentially unchanging physical laws. During a steady rain, for example, we can observe rills being carved in bare earth as the runoff turns muddy (fig. 1.3). Although it may be difficult to see the individual particles being loosened and carried away, we can easily measure the changes after the process has continued for a while. In one day, for instance, an inch-deep gully may have been formed. We can then estimate the probable

Figure 1.4. Gully 2 feet deep

Figure 1.3. Erosion caused by runoff

changes over a longer period. If it were to rain only twice a year, the gully would be 2 inches deep after 1 year, 2 feet deep in 12 years (fig. 1.4), 200 feet deep in 12 centuries—and after only 30,000 years, as deep as the Grand Canyon (fig. 1.5).

It took much longer than 30,000 years to carve the Grand Canyon out of solid rock. But solid rock, like soil, is worn away by natural processes— more slowly, but just as surely. In fact, the ancestral Colorado River began cutting the Grand Canyon when the Kaibab Plateau began rising about 9 million years ago. To carve it to its present 1-mile depth, the river cut downward a mere two-thirds of an inch per century—a trivial task for the mighty Colorado.

Figure 1.5. Grand Canyon (Courtesy of J. R. Balsley, United States Geological Survey)

4

The principle of uniformitarianism applies to all other geologic processes known to be operating today. A band of cross-bedded sandstone exposed in a cliff (fig. 1.6) shows what can happen to desert sand dunes. An earthquake may leave fresh fault scarps running miles across the landscape as direct evidence that mountains are still growing (fig. 1.7). A river that must be remapped every few years illustrates how both erosion and deposition of water-borne

Figure 1.7. Fault scarp caused by an earthquake (Courtesy of J. R. Stacy, United States Geological Survey)

Figure 1.6. Cross-bedded sandstone in Canyon de Chelly (Photo by J. Morris)

sediments occur continually (fig. 1.8). Oceans, rivers, glaciers, volcanoes—all have left evidence of their effects far back in the geologic record.

The rate and magnitude of geologic processes may change over time. Some changes are permanent: the earth is cooler and its crust thicker than in the geologic past. Others are cyclical, such as the building up and wearing down of mountains. Even a volcanic eruption is a geologically minor event; thousands of such eruptions occurred over many millions of years to give the Cascade Range its present form.

In any given location, different processes dominate at different times. When mountains are rising, erosion outpaces deposition; when land subsides beneath the sea, deposition takes over. Some phenomena, such as the impact of a large asteroid, may happen so seldom as to have no counterpart in recent geologic history; but the effects of forces great and small upon the structure of the earth can be traced back almost to the beginning of geologic time.

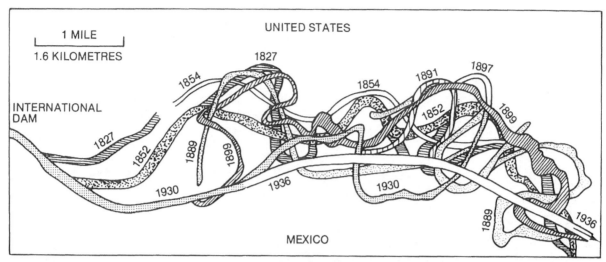

Figure 1.8. Shifting channel of Rio Grande near El Paso (From *Physical Geology* by Robert J. Foster. Copyright 1983 by Charles E. Merrill Publishing Company. Reprinted by permission of the publisher.)

GEOLOGIC TIME

When geology was dominated by the concepts of special creation and catastrophism, geologic time was of little consequence. The conventional wisdom of the Middle Ages was, as Shakespeare wrote in *As You Like It*, that "the poor world is almost 6,000 years old." Even in Shakespeare's time, however, there was evidence that the planet was much older.

Scientists had long been aware of the mystery of fossils, which were apparently the remains of plants and animals found in solid rock (fig. 1.9). How could a fish that lived in the ocean end up on a mountaintop? Some believed that fossils were not the remains of life forms but strange manifestations of the rocks themselves that simply happened to resemble plants or animals. Others believed them to be the remains of organisms drowned or stranded during Noah's flood.

Those who embraced the inorganic theory had difficulty explaining why fossils often duplicated the finest details of living organisms. Could this happen by sheer coincidence? It seemed unlikely. Organic theorists, on the other hand, could not explain how the carcasses of fish, birds, and plants could have been left deep within a mountain during a flood. Why did they occur in such distinct layers? Fossils found in the uppermost layers resembled living organisms; deeper layers contained many that had never been seen alive. Certain ancient fossils found in England could be

Figure 1.9. Fossils (Courtesy of W. T. Lee, United States Geological Survey)

found in Europe in the same type of rock—and often on the other side of the Atlantic as well.

Many geologists recognized that sedimentary rocks were composed of particles of older rock that had been deposited by running water or other agents and then consolidated into new rock. Nicolaus Steno, a physician and naturalist who lived in Florence, Italy, showed in 1669 that layers of sedimentary rock were deposited atop one another in such a way that the youngest layers occurred on top and the oldest on bottom. This simple, logical, and (in retrospect) obvious idea implied that the fossils occurring within the layers were the same age as the layers themselves. By observing the processes going on around them, scientists could estimate how long it would take to deposit a given thickness of sediment—and by extension, the age of the earth, with its thousands of feet of sedimentary rock.

Geologists began to see that the earth had been adding layers to itself for much longer than 6,000 years. One nineteenth-century estimate of the earth's age, based on the time it would take for freshwater streams to bring in the salt now present in the world's oceans, was 100 million years. As was to be shown later, however, even 100 million years was but a small fraction of the earth's true age.

Physicists in the twentieth century discovered that radioactive elements, present in trace amounts in most rocks, lose their radioactivity over time and become transformed into nonradioactive elements. The rate of decay for a particular radioactive element is proportional to the number of atoms of the element present. Half of these atoms will decay during a period called the *half-life*, which is different for each radioactive element. For instance, radioactive uranium-235, with a half-life of 713 million years, decays to nonradioactive lead-207 at a rate such that after 713 million years, half the original U-235 present will have decayed to Pb-207, and after another 713 million years, half of the remaining U-235 will have decayed, and so forth (fig. 1.10).

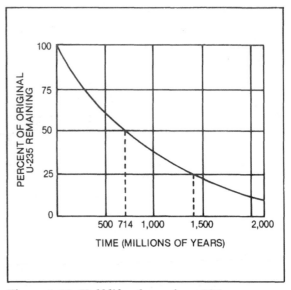

Figure 1.10. **Half-life of uranium-235**

In nature, lead is a mixture of isotopes, including lead-207 and others. The rock samples in which radioactive uranium-235 has decayed will have higher proportions of lead-207 in them than the other lead isotopes, and the ratio increases with time at a known rate. By measuring the ratios of different lead isotopes in a rock sample, a physicist can estimate with some accuracy how old the rock is.

Using the uranium-lead method, scientists have estimated that certain ancient rocks in western Greenland are 3.8 billion years old. Many meteorites have been found to be 4.6 billion years old, as have the oldest samples of moon rocks brought back by our astronauts. Geologists believe that the earth was formed at the same time as the rest of the solar system—about 4.6 billion years ago, according to the best evidence—and that 800 million years passed before the oldest rocks now found on the surface were formed.

During the past 200 years, geologists have developed a system of classifying rocks by relative age (fig. 1.11). These classical divisions of geologic time are continually updated and refined using radiometric estimates of the rocks' ages. The oldest rocks are those of the **Precambrian** era, a

ERA	PERIOD	EPOCH	AGE (MM. Yrs)	EVENTS
CENOZOIC	QUATERNARY	HOLOCENE	0.01	End of most recent ice advance in North America
		PLEISTOCENE	1	Recent ice ages
	TERTIARY	PLIOCENE		Colorado River begins cutting Grand Canyon
		MIOCENE		India collides with Asia
		OLIGOCENE		
		EOCENE		
		PALEOCENE	65	Rocky Mountains formed Mammals proliferate Dinosaurs extinct
MESOZOIC	CRETACEOUS			First flowering plants
			135	
	JURASSIC			First birds
			190	
	TRIASSIC			Atlantic rift opens First mammals First dinosaurs
			220	
PALEOZOIC	PERMIAN		280	Glaciation of southern Gondwanaland First reptiles
	PENNSYLVANIAN		325	
	MISSISSIPPIAN		350	Africa collides with North America (Appalachians form)
	DEVONIAN		400	First land animals (amphibians) First rooted land plants
	SILURIAN		430	Glaciation
	ORDOVICIAN		500	First vertebrates (fish)
	CAMBRIAN		600	
PRECAMBRIAN				First complex animals (worms) First green algae
			2000	World iron ore deposits forming
			3000	Oldest evidence of life
			3800	Oldest known rocks
			4600	Formation of planets, solar system

Figure 1.11. Geologic time

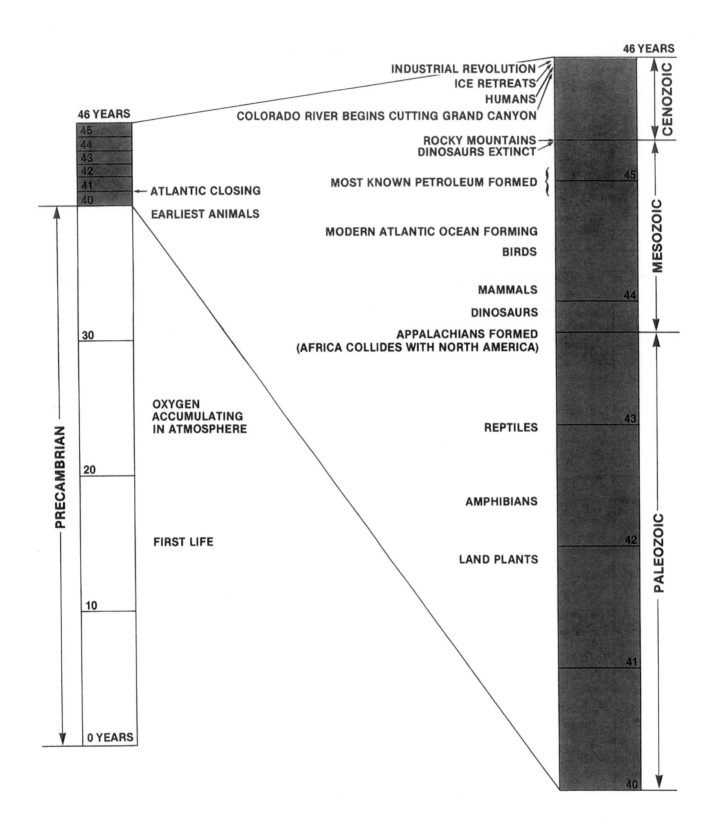

Figure 1.12. Life span of Terra

span of 4 billion years from the earth's beginning until 600 million years ago. This era encompasses most of the earth's existence; yet little is known about it, because much of the rock that was formed in the Precambrian era has been greatly altered by mountain-building forces. There were no plants or animals on the land, no fish in the sea. The only fossils found in Precambrian rocks are of primitive life forms such as bacteria and blue-green algae that lived in shallow coastal waters. After photosynthetic algae became abundant and began using carbon dioxide and releasing free oxygen, the atmosphere began to evolve toward its present nitrogen-oxygen composition.

The **Paleozoic** era, from 600 million to 230 million years ago, witnessed the first great diversification of life forms. Marine invertebrates developed hard shells; invertebrates with exoskeletons evolved into the first insects; other hard-shelled forms developed spinal cords and became the first bony fishes. Plant life began to flourish on dry land, followed by the first air-breathing animals—scorpions. Amphibians came along in the Devonian period, and their descendants, the reptiles, in the Pennsylvanian. The oldest surviving deposits of petroleum began as accumulations of organic detritus in shallow inland seas and along ocean margins.

The **Mesozoic** era, 230 to 65 million years ago, was the time of the dinosaurs. The first mammals also appeared but, overshadowed by the large and diverse "thunder lizards," were represented by only a few primitive forms. Birds and flowering plants appeared. All land was grouped together in one great "supercontinent," which began to break apart into the continents of today. In the Cretaceous period, 135 to 65 million years ago, a great tropical ocean lay where the Middle East is now; more than half of the world's known petroleum accumulated along the margins of this ocean during a brief (30-million-year) interval.

Cenozoic events (those during the last 65 million years) included the mysteriously sudden extinction of the dinosaurs and many other species on land and at sea; the raising of the Rockies, Alps, and Himalayas; and most recently, the advances and retreats of a succession of ice sheets in North America and Europe.

Trying to comprehend 4.6 billion years is like trying to visualize a road 4.6 billion miles long: it is beyond understanding because it is far beyond our experience. If you divided a highway from New York to San Francisco into 4.6 billion parts, each part would be only ⅟₂₅ inch long—about a millimetre. It is easier to compare 4.6 billion years with a more familiar time span—such as the life of a person 46 years of age, who for want of a more appropriate name shall be called Terra. Using Terra's life scale, we can get a better idea of how the major events of geologic time are related to each other (fig. 1.12).

Terra's early childhood is a mystery to us: before she was 8, she tore the pages out of her rock diary as fast as she could write them. She was 15 before her oceans began showing signs of life—single-celled organisms. In her 20s she began replacing the carbon dioxide in her atmosphere with oxygen.

Her life became more interesting to us when she turned 40—just 6 years ago—and the first recognizable animals appeared. Between her 42nd and 43rd birthdays, land plants, amphibians, and reptiles came along. She was nearly 44 when the Appalachians rose and dinosaurs began to roam.

Most of her petroleum was formed about a year ago. Eight months ago dinosaurs became extinct and the Rocky Mountains began rising; last month the Colorado River started cutting the Grand Canyon across a featureless plain. Fifteen days ago humans with stone tools kicked off their campaign for dominance. Fifty-five minutes ago the ice retreated for the last time, opening the northern continents to nomadic hunters.

And 58 seconds ago the Industrial Revolution led to the draining of most of Terra's supply of petroleum, which she had been saving up since she was 40.

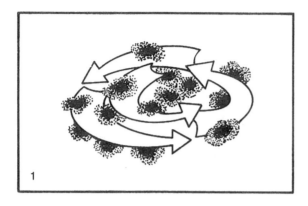

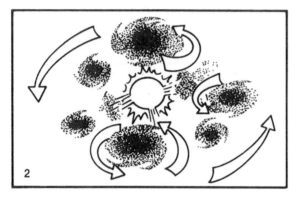

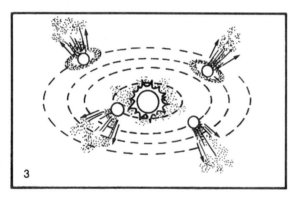

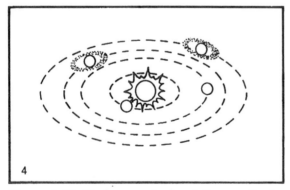

Figure 1.13. Modern protoplanet theory of solar system formation

HOW THE EARTH BEGAN

The earth's present structure can best be understood as the result of earth processes continued over a very long period. But how did the earth start? What did it look like when it was first formed? These questions are hard to answer because all direct evidence of the earth's formation has been obliterated by the changes that have occurred since.

Current theories hold that the solar system developed from a cold, turbulent cloud of particles made up of all the elements now found in the earth and the sun (fig. 1.13). These particles, containing not only the heaviest metals but also the lightest gases, drew together in local aggregations by mutual gravitational attraction. As the slowly-rotating cloud contracted, gravitational compression heated the largest aggregation, in the center, hot enough to start a nuclear fusion reaction. The newborn star began "burning" its hydrogen into helium, meanwhile pouring out a constant stream of particles. This "solar wind" began driving off the lighter, more volatile elements from the nearest parts of the cloud, where smaller clusters of particles were forming into planets. The inner planets—Mercury, Venus, Earth, and Mars—were washed more intensely by the solar wind and lost more of their gases than the outer planets—Jupiter, Saturn, Uranus, and Neptune, the "gas giants." Saturn, with its gas envelope and its rings of ice and rock particles, hints at what the early planetary aggregations may have looked like.

The heavier constituents of the earth's cloud, stripped of their gaseous envelope, fell together about 4.6 billion years ago. Due to radioactivity and gravitational compression, the interior became hot enough to melt. Denser elements such as iron and nickel collected at the center to form the *core* (fig. 1.14). Lighter elements such as magnesium, aluminum, and silicon formed the solid outer layers—the plastic *mantle* and the rigid *crust*. The *outer core* remained fluid, but pressure at the center caused the iron and nickel in the *inner core* to

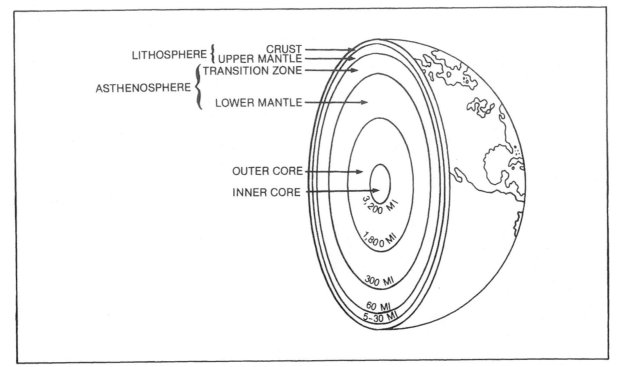

Figure 1.14. Cutaway view of Earth

resolidify. The metallic, part-solid and part-liquid core began to act as a giant electric generator, creating a magnetic field that extended well beyond the planet's surface.

Of the lighter elements that form the crust, oxygen, silicon, and aluminum constitute about 82% by weight, mostly in the form of silicon dioxide and aluminum silicates. Next in abundance is iron, the main element of the earth's core and of the earth as a whole. Carbon, the essential element of living matter and of fossil fuels, constitutes only a small fraction of the earth's crust. It was present at first mostly as atmospheric methane. Later this methane was converted to carbon dioxide and water vapor as photosynthetic plants released an abundance of oxygen into the atmosphere.

Earth scientists still do not know whether the entire planet became molten shortly after it fell together or whether it warmed gradually and melted from within, leaving the surface cool and brittle. Whichever explanation is correct, the crust was at first thin and unstable. Most geologists now believe that the earth had no surface water at first, and that both free water and water chemically bound up in the rocks reached the surface in volcanic eruptions and then cooled and condensed to form the first oceans.

For the first two billion years or so the only land visible above the surface of these shallow oceans was in the form of many large and small islands. Although there were many small volcanoes, the crust was too thin and weak to support large mountains. Bacteria and primitive algae had arisen from a complex of chemicals in the sunlit upper layers of the ocean. Around 3 billion years ago some communities of these organisms evolved into large, matlike colonies in the shallow sea margins. Other forms developed the ability to use the energy of sunlight to make simple sugars from carbon dioxide and water by the process of *photosynthesis*.

After 2.5 billion years, the crust had become thick enough and stable enough for some land areas to stand well above sea level. One result was an increase in erosion and the deposition of large quantities of sediments in low-lying areas and on the submerged edges of the continents. These early sedimentary basins accumulated thick sequences of sandstones, shale, and limestone.

The first high mountains were built, paradoxically, along these humble continental margins. Titanic forces raised and buckled and deformed the thick sedimentary beds of the continental shelves and transformed them into mighty mountain ranges. Mountain building occurred in episodes, with stable, quiet periods in between during which erosion became dominant and wore the mountains down. The cause of these periodic upheavals has been the subject of speculation and debate since geology became a science. Only recently have geologists formulated a coherent

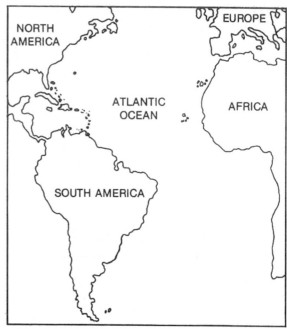

Figure 1.15. Atlantic coastlines

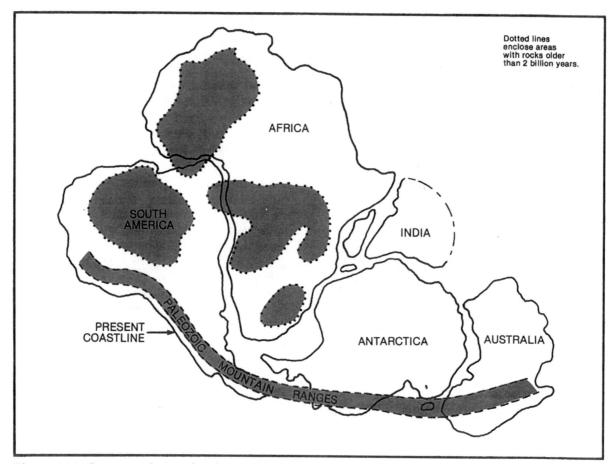

Figure 1.16. Correspondence of rock types and structural trends in southern continents

theory that accounts for most of the forces that shape the earth.

PLATE TECTONICS

Ever since the first accurate maps of the New World were drawn, geographers have noted the similarity of the eastern and western margins of the Atlantic Ocean (fig. 1.15). By the middle of the nineteenth century some observers were speculating that the land masses on either side had once been joined in a single continent. In 1910, Alfred Wegener, a German meteorologist, theorized that the continents had migrated across the ocean floor like rafts drifting at sea. He called his theory *continental drift*.

Wegener had an impressive array of facts in support of his outlandish theory. Geologists in South America and South Africa had amassed evidence that the rock types, minerals, and structural trends along the Atlantic coasts of both continents matched like patterns on a jigsaw puzzle (fig. 1.16). Other observations indicated that parts of South America, South Africa, Madagascar, India, Australia, and Antarctica had once been covered by a single glacial ice cap (fig. 1.17). Fossil remains of extinct land animals found in isolated areas on opposite sides of oceans could be explained in only two ways: either there had been "land bridges" thousands of miles long that had vanished without a trace, or continents that had once been joined together had drifted apart.

Other observations supported the idea that continents had traveled great distances. Antarctica has great seams of coal that could only have originated in a warm climate quite unlike its present deep freeze—that is, when the continent was closer to the

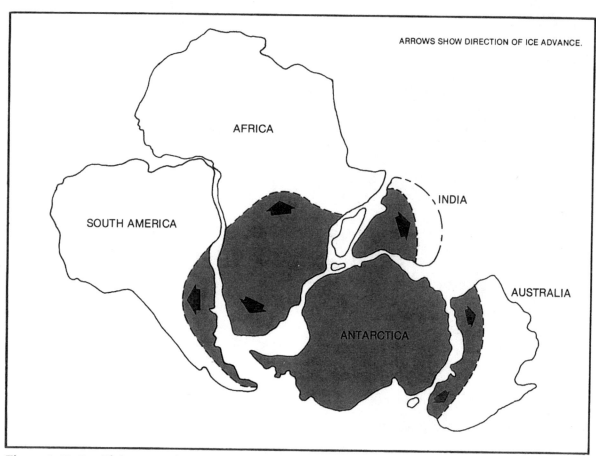

Figure 1.17. Pre-Pleistocene glaciation of southern continents

equator. Similarly, evidence of ancient ice sheets in the Sahara desert could only have been produced at latitudes closer to one of the poles. But how could continents of rigid granite move across ocean beds of solid rock?

As it turned out, Wegener was right in principle but wrong in detail. The continents don't plow through the crust like ships through water, but ride along on great moving crustal plates like boulders on a giant conveyor belt. The earth's crust today is divided into six major plates, and several smaller ones, atop which the continents are carried away from a system of midocean

ridges and toward another system of deep-sea trenches. This theory, which has explained most of the mysteries that confounded both Wegener and earlier geologists, is called *plate tectonics,* after the Greek word for "builder," *tektonikos.*

The key to the puzzle was found in the oceans. In the 1950s, geophysicists studying the ocean floor found that the Mid-Atlantic Ridge, a submerged volcanic mountain chain winding almost exactly down the center of the ocean, was flanked by a symmetrical pattern of parallel bands of rock magnetized in opposite directions (fig. 1.18). Scientists already knew that the earth's

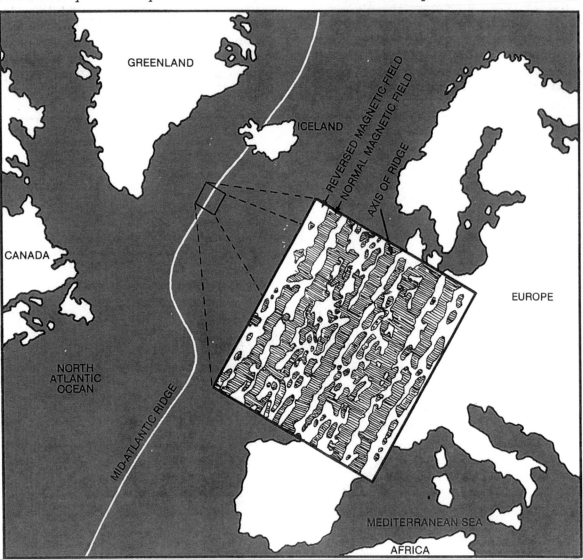

Figure 1.18. Seafloor magnetic anomalies along Mid-Atlantic Ridge

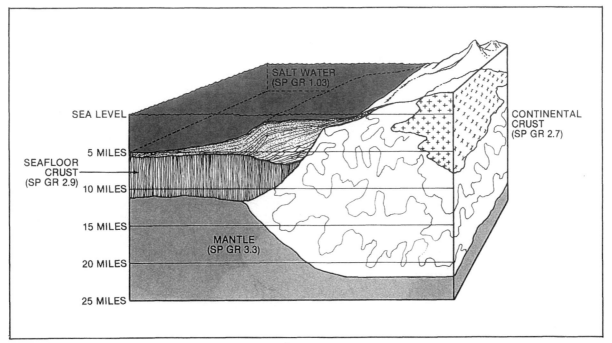

Figure 1.19. Isostasy of seafloor and continental crust

magnetic field reverses its polarity at erratic intervals of from 50,000 to 3,000,000 years – on the average, every 450,000 years. The seafloor magnetic anomalies proved that molten rock rising in the center of the ridge solidifies and spreads outward in both directions, forming a "graph" of the earth's magnetic reversals in parallel bands on the seafloor.

The Mobile Crust

From local measurements of gravitational strength, geologists know that the continents are composed of lighter rocks than the seafloor. Although they cannot migrate through the rigid oceanic crust, masses of granite "float" like icebergs upon the dense, semimolten rock of the upper mantle: the higher the mountain, the deeper its roots (fig. 1.19). This principle is called *isostasy*.

The early crustal plates, thin and unstable, moved more rapidly than the thick plates of today and were affected over much broader areas by events occurring at their margins. By 2.5 billion years ago,

however, they probably began to resemble much more closely their present-day counterparts. Driven by slowly diverging convection currents in the semisolid, deformable mantle, a large land mass would separate along a *rift zone*. As the rift widened and the land masses on both sides moved apart, new oceanic crust would be formed. The two "daughter' continents were thus pushed or carried away from the ridge on the spreading seafloor. Since the Atlantic began opening about 200 million years ago, North and South America have been moving away from Europe and Africa at a brisk 1¼ inches per year.

Geologists today know approximately what the world looked like 200 million years ago, before the Atlantic Ocean was born (fig. 1.20A). All land was grouped into one mass that geologists now call *Pangaea*, from the Greek words meaning "all the world." Notice the triangular gulf between Asia and Africa. This warm, tropical sea, which geologists now refer to as the *Tethys Sea*, accumulated thick deposits of sediments from the land and supported an

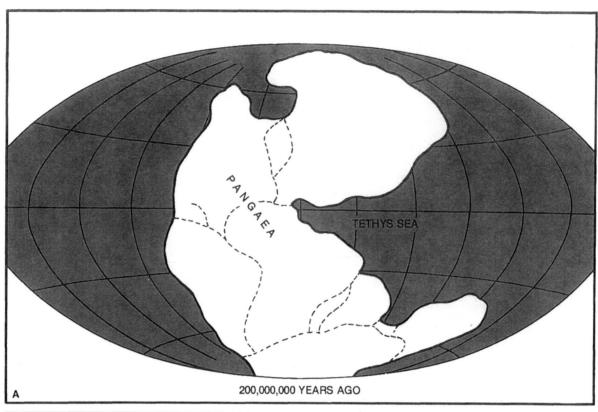

A   200,000,000 YEARS AGO

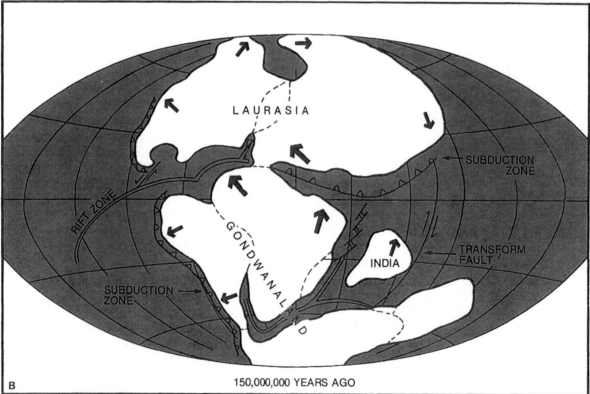

B   150,000,000 YEARS AGO

Figure 1.20. Breakup of Pangaea into modern continents

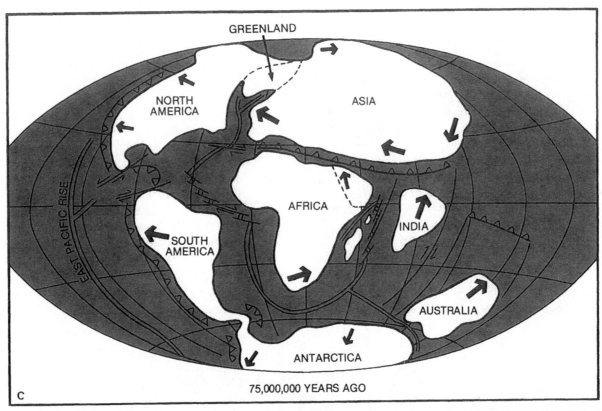

C   75,000,000 YEARS AGO

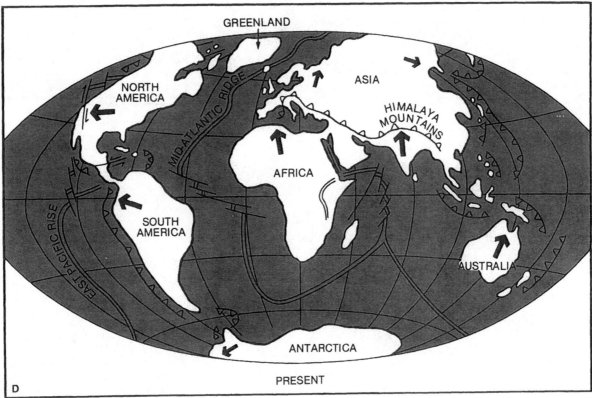

D   PRESENT

Figure 1.20.—*Continued*

abundance of marine life. More than half the world's known petroleum is found along its former shores.

About 150 million years ago, the North Atlantic had started to open up between North America and Africa, and the South Atlantic between the southern tips of Africa and South America (fig. 1.20B). The Tethys Sea began closing as Africa and Asia pivoted toward each other. A triangular fragment – India – separated from the southern lands and raced north. Volcanoes grew along the western coasts of the Americas as the continents overrode the Pacific Ocean floor.

By 75 million years ago, the world had begun to look more like today's familiar globe (fig. 1.20C). Greenland still bridged the North Atlantic, but the South Atlantic

was open from the equator to the Antarctic. The Tethys Sea had been all but squeezed out. India was nearing violent collision with Asia. The seafloor rift south of India had grown eastward to separate Australia from Antarctica.

The world today shows clearly the results of the last 200 years of plate motion (fig. 1.20D). India has collided with mainland Asia but has not yet stopped moving; the Himalayas are still rising. The young Red Sea Rift is extending itself through eastern Africa. The East Pacific Rise, a rift zone that connects through the Indian Ocean with the Mid-Atlantic Ridge, has been overridden by the west-moving North American plate – a situation by which California has gained dubious distinction.

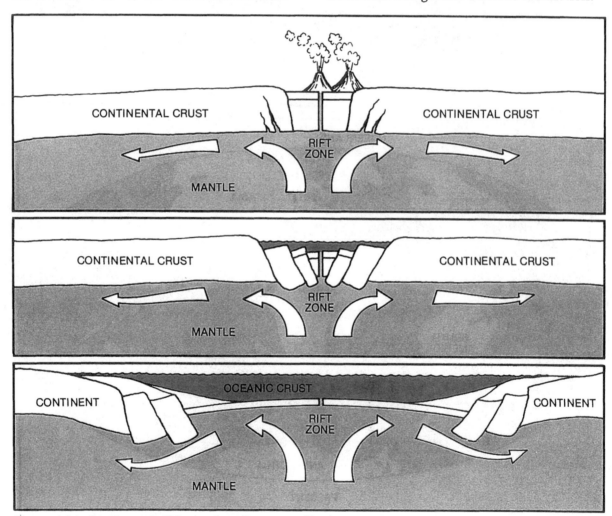

Figure 1.21. Continental rifting and seafloor spreading

Rifting

One consequence of the thousands of miles of horizontal plate motion is thousands of feet of vertical adjustment. The region in which a rift begins to form is uplifted; after separation of the continental masses, the rift zone itself remains higher than the seafloor on either side (fig. 1.21). The seafloor sinks as it moves away from the ridge, and the trailing edge of the continent sinks with it. Wherever the continental shelf is submerged, it becomes blanketed with layers of sediment from the land.

When a rift begins to develop across continental crust, as is happening in East Africa, the continental crust comes under tension. This stress often results in the formation of basins known as *grabens*—blocks of crust that have subsided relative to adjacent blocks (*horsts*). Grabens on the continental shelf often fill with shelf sediments that can trap oil, as in the North Sea. A rift that fails to develop into an ocean basin may end up as a narrow zone of grabens filled with oil-generating and oil-trapping sediments; southern Oklahoma's Anadarko Basin is an example.

Geologists don't yet know why the continents move. Some theorize that slow convection currents of hot, semimolten rock originating deep within the mantle rise to the surface and spread horizontally beneath the ocean floor, pushing the crust apart along the midocean ridges. Others hold that cooler material sinking back into the lower mantle pulls the plates along and downward into the deep trenches, and that molten rock rises into the midocean rifts only to fill the gap left at the trailing edges. It now seems likely that both mechanisms are involved to some degree and that the driving mechanism of the mobile crust may ultimately prove to be as complex as atmospheric or oceanic circulation.

Transform Faulting

A *fault* is a break along which two blocks of the crust slide past one another. As a consequence of relative plate motion, the ocean floor is crisscrossed with faults. Along the edges parallel to its motion, a plate slides past its neighbor with very little interaction. These plate boundaries, most of which occur on the ocean floor, are called *transform faults*.

California's notorious San Andreas Fault is one of the few examples of transform faulting on land (fig. 1.22). Thirty million years ago, a transform fault across the East Pacific Rise was overridden by the west-moving North American plate. After this fault passed beneath the edge of the continent, the northwest-moving Pacific plate wrenched loose a piece of the west coast. This fragment has become attached to the Pacific plate and has slipped hundreds of miles north, with its northern tip near San Francisco today. It has moved not smoothly but in a series of sudden, devastating earthquakes at intervals of many years. If the movement continues, Los Angeles will slide beneath Alaska 80 million years from now.

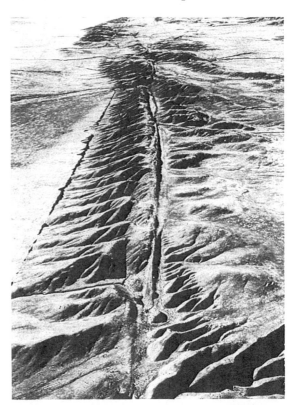

Figure 1.22. San Andreas Fault in California (Courtesy of R. E. Wallace, United States Geological Survey)

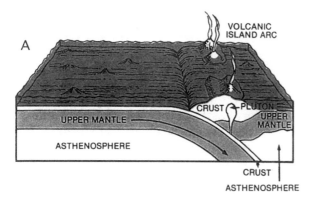

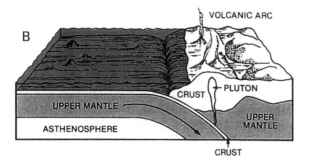

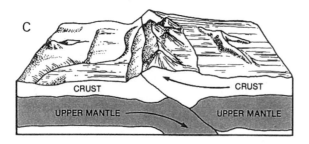

Figure 1.23. Plate convergence

Convergence

The most spectacular geologic events occur at the leading edge of a moving plate. As one plate impinges upon the edge of an adjoining plate, something has to give. The three principal types of plate convergence are ocean to ocean, ocean to continent, and continent to continent.

When oceanic crust meets oceanic crust, one plate dives beneath the other (fig. 1.23A). A deep *trench* is formed in the ocean floor along the line of convergence, which is called a *subduction zone*. The plate that descends into the hot mantle is partially melted, its lighter fractions rising through fissures in the heavier, unmelted crust above. At the surface, the molten rock erupts in a line of volcanoes, eventually forming an *island arc* parallel to the trench. The Windward Islands of the Caribbean are an example of a young island arc; Japan and the Philippines are older arcs. These small land masses may grow and merge or collide with larger continental land masses. Much of the west coast of North America is believed to have grown in this way; recent evidence indicates that parts of the south coast of Alaska were once islands in the South Pacific.

When oceanic crust collides with a continent, the lighter continental crust overrides the ocean floor (fig. 1.23B). The descending plate melts fractionally, often producing a *volcanic arc* along the edge of the continent that is underlain by *plutons,* large subterranean bodies of slowly cooling igneous rock. Erosion may later expose these bodies, as in the present-day Sierra Nevadas in California.

When two continents collide, great mountain ranges such as the Alps and the Himalayas are pushed up (fig. 1.23C). Thick continental crust does not behave like thin, heavy oceanic crust; instead of sliding down into the mantle, it crumples, folding and breaking into huge slabs (*nappes*) that pile atop one another like wind-driven ice floes in the Arctic. In these massive folded and overthrust belts, much of the rock may be changed by heat and pressure into new forms. This process is called *metamorphism.*

The rate at which mountains are raised is a good illustration of the principle of gradualism. Although in the geologic sense the collision of two continents is a sudden, catastrophic event, the changes observable in a human lifetime may seem insignificant. For example, the rock on one side of a fault might rise one foot in a few seconds during a major earthquake, only to lose 9 inches by erosion in the hundred years before the next quake. Yet a sedate 3 inches per century can build a 2-mile-high range of mountains in 4 million years—a mere geologic eyeblink.

Mountain ranges build down as well as up. As the crust piles up, the additional weight forces the mantle deeper, just as loading a boat makes it ride deeper. This isostatic adjustment may pull down nearby crust, forming basins well above sea level where sediments eroding from the mountains can accumulate.

ROCKS AND MINERALS

A *mineral* is a naturally occurring inorganic crystalline element or compound. Although some rocks, such as rock salt, are made up of grains of a single mineral, most rocks are assemblages of different minerals.

A mineral has a definite chemical composition and characteristic physical properties such as crystal shape, melting point, color, and hardness. However, most minerals, as found in rocks, are not pure. *Quartz,* a form of silica, has a definite chemical formula (SiO_2) but usually contains impurities that give it color. *Feldspar,* on the other hand, includes within its chemical range a wide variety of potassium, sodium, and calcium aluminum silicates. Two common varieties are *orthoclase,* which is mostly $KAlSi_3O_8$, and *plagioclase,* a combination of $NaAlSi_3O_8$ and $CaAl_2Si_2O_8$. Varieties of feldspar make up about half the rocks of the earth's surface.

Because minerals vary in solubility, hardness, and other properties, they tend to occur in different proportions in different environments. Consider, for example, a rock that contains equal proportions of *silica* (SiO_2) and *mica* $[KAl_3Si_3O_{10}(OH)_2]$. Silica is relatively hard and insoluble, while mica is physically fragile and easily decomposed into other minerals. As the rock containing both is weathered and eroded, the mica is carried only a short distance before being transformed into clay, but the grains of silica persist. Therefore, sediments derived from this rock will contain more silica than mica.

If you examine a typical rock, you will see that it is composed of individual mineral grains of various types. In some rocks the grains are so small that they can be seen only under a microscope; in others, they are quite large. *Texture,* the size and arrangement of these mineral particles, is one of the principal descriptive properties geologists use to classify rocks.

Two basic kinds of texture are crystalline and clastic. In rocks with *clastic* texture, the grains, which are broken fragments of rocks, minerals, or organic debris, may touch each other but are mostly surrounded and held together by a cement such as calcite (fig. 1.24). Often the spaces between

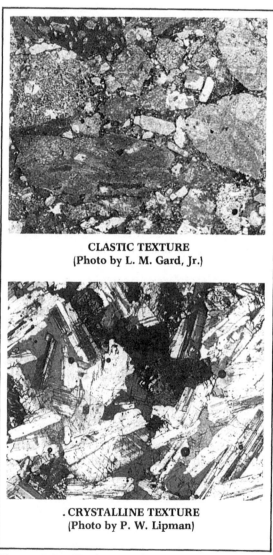

CLASTIC TEXTURE
(Photo by L. M. Gard, Jr.)

CRYSTALLINE TEXTURE
(Photo by P. W. Lipman)

Figure 1.24. Rock texture (Courtesy of United States Geological Survey)

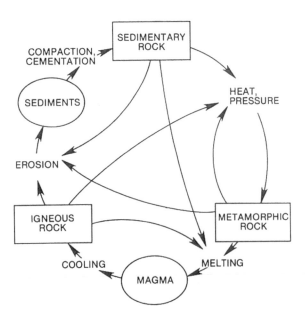

Figure 1.25. The rock cycle

the grains are not completely filled, leaving openings or voids that can contain water or other fluids. In *crystalline* texture, the grains form and grow from the molten rock material as it solidifies, and so they interlock or touch each other on all faces. Minerals can also crystallize from groundwater solutions within existing rock. Rocks with crystalline texture usually have very low porosity.

Primary rock classification is by origin: *igneous*, *sedimentary*, and *metamorphic*. Rock in any one of these categories can be transformed by natural processes into its counterpart in another category. The relationships of these rock types and processes is shown diagrammatically as the *rock cycle* (fig. 1.25).

Igneous Rock

Igneous rocks are those formed directly from molten rock material, or *magma*. When the planet was formed, the original crust was entirely igneous; today the percentage is 65%. Two principal types of igneous rock are *intrusive* (or *plutonic*), those that have solidified below the surface, and *extrusive* (*volcanic*), those that have formed on the surface.

Magma that cools underground loses heat slowly. The more slowly it cools, the more its individual minerals tend to separate before crystallizing and thus form larger crystals. *Granite*, the most common intrusive igneous rock, has crystals that are easily seen by the unaided eye (fig. 1.26). A rock that has crystallized over a longer period may have crystals measured in inches or feet. The most common types of plutonic rock are *gabbro*, which solidifies at high temperatures; *diorite;* and *granite*, a relatively low-temperature rock.

Magma that reaches the surface is called *lava;* extrusive igneous rock is hardened lava from volcanic eruptions. Because it has lost its heat rapidly to the atmosphere, its grains are usually smaller than those of intrusive rock and may be visible only under a microscope. Some lava cools so rapidly that it does not form crystals at all; the result is a volcanic glass known as *obsidian*. The most common extrusive rocks are *basalt, andesite,* and *rhyolite*, which correspond to the plutonic rocks listed above.

Since igneous rocks form from a cooling body of magma or lava, they are usually crystalline and nonporous. However, *pyroclastics*, composed of fragments of ash from volcanic explosions, have clastic texture and are porous. Gas-filled lava that has cooled rapidly may form a vesicular type of obsidian called *pumice*, the rock that floats on water.

Sedimentary Rock

When any type of rock is exposed at the surface, it becomes subject to weathering and erosion. *Weathering* processes are those that break down the structure of the rock by chemical and physical attack. *Erosion* is the removal of weathered rock or soil particles by flowing water, wind, moving ice, or other agents. When weathering has proceeded far enough, an erosional process may complete the job of separating the particles from the parent rock.

The rock and soil particles carried away by erosion eventually come to rest in a sedimentary deposit, often far from their

GRANITE
(Photo by A. Z. Hiltanen)

OBSIDIAN
(Photo by C. Milton)

LAVA
(Photo by D. A. Swanson)

Figure 1.26. Igneous rocks (Courtesy of United States Geological Survey)

source. The largest and heaviest particles, requiring the most energy to transport, are the first to settle out, followed by particles of decreasing size. As a result, fine silt and clay particles are carried farthest from their origin and deposited where the moving water loses the last of its flow energy. Eroded particles may eventually be consolidated as sedimentary rocks (fig. 1.27).

Although they comprise only 8% of the volume of the earth's crust, sedimentary rocks cover 75% of the land surface. Most sedimentary rocks are porous and therefore capable of containing fluids. The fact that most petroleum accumulates in sedimentary rocks makes them of prime interest to the petroleum geologist.

Figure 1.27. Sedimentary rocks (Courtesy of G. K. Gilbert, United States Geological Survey)

Figure 1.28. Metamorphic rock (Courtesy of J. Gilluly, United States Geological Survey)

Metamorphic Rock

Any rock that has been changed by pressure and heat while in the solid phase is termed *metamorphic* (fig. 1.28). When *shale*, a common sedimentary rock, undergoes deep burial, heat and pressure fuse individual mineral grains into a metamorphic rock known as *slate;* more intense metamorphism produces *schist*. Similarly, limestone becomes *marble.* The mineral composition of rocks undergoing low-grade metamorphism usually does not change significantly; under greater pressures and temperatures, however, some minerals may become chemically altered, and fractions of the rock may even melt, recombine, and recrystallize. One type of low-grade metamorphism, the alteration of limestone to dolomite, involves a chemical replacement of calcium by magnesium in solution.

Metamorphism always results in crystalline texture; either new crystals are formed from the elements present in the original rock, or the existing grains are deformed and molded into an interlocking structure. Metamorphic rocks usually have little or no porosity.

Two basic kinds of metamorphism are *contact* and *regional* (fig. 1.29). Contact metamorphism occurs when an intruded body of molten igneous rock changes the rocks immediately around it, primarily by heating and by chemical alteration. Regional metamorphism, much more common and affecting much larger areas, occurs in bodies of rock that have been deeply buried or greatly deformed by tectonic changes; these changes are caused by heat and pressure.

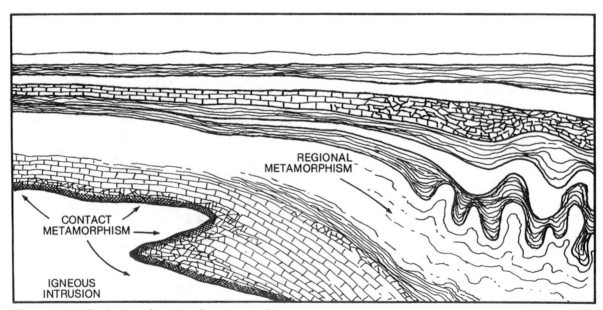

Figure 1.29. Contact and regional metamorphism

Metamorphic rocks are usually grouped into two classes: *foliated* metamorphics, which have a layered look that is not necessarily associated with the original layering in sedimentary rocks; and *nonfoliated* metamorphics, which appear massive and homogeneous. The final form and appearance of any metamorphic rock is a product of its original composition and the type and intensity of the metamorphic forces involved. About 27% of the earth's crust is composed of metamorphic rocks.

It is the crust, of course, that most concerns us as human beings and as finders and users of oil and gas. The crust is different from the molten interior of the earth; it is the cool skin where life exists. Its heights are worn down by the constant wash of weather; its basins are filled with water and rock particles and the remains of once-living organisms. In the search for oil, the petroleum geologist concentrates on these basins, where sediments and organic matter accumulate in a sort of giant pressure cooker.

2

SEDIMENTATION

Sedimentary rock is the most interesting type of rock for the petroleum geologist because that is where nearly all the world's petroleum is found. To understand this phenomenon, we must learn more about sedimentation—how sedimentary particles are formed, transported, deposited, and transformed into the great sheets of rock that cover most of the world's land area.

ORIGINS OF SEDIMENTARY PARTICLES

How sedimentary rock is formed can perhaps best be demonstrated by examining its smallest unit—the sedimentary particle. As does the rock as a whole, the individual particle embodies the history of both its source material and the changes it undergoes on the earth's surface.

Clastics

Sediments are classified primarily by grain size (table 2.1). Gravel, sand, and silt particles can be of a variety of minerals— quartz and feldspar are common—while clay particles are microscopic platelets of various hydrous aluminum silicates. Gravel, sand, and silt are mostly non-cohesive; that is, they do not stick together. Clay, on the other hand, is very cohesive; its particles are attracted to one another by minute electrical charges and adsorb water readily, causing clay to swell.

Most sedimentary particles are formed by the breakdown of large rock masses into smaller pieces by climatological processes. These sediments are known as *clastics*. Climate accomplishes this breakdown in two principal ways—physically and chemically. Different kinds of rock are affected in

TABLE 2.1

CLASSIFICATION OF SEDIMENTS BY GRAIN SIZE

Particle Name	Diameter Range	Rock Type
Gravel	Larger than 2 mm	Conglomerate
Sand	$1/16$ mm–2 mm	Sandstone
Silt	$1/256$ mm–$1/16$ mm	Siltstone
Clay	Smaller than $1/256$ mm	Shale

different ways by these processes, which together are known as *weathering*.

Physical weathering. Sedimentary particles are derived from rocks formed deep underground and then uplifted and exposed on the surface. With the weight of the overburden removed, internal stresses often cause such a rock mass to expand and develop cracks and fissures – that is, *joints*. These joints grow as the exposed rock expands and contracts in response to changes in air temperature. Intersecting sets of parallel joints may divide the rock into square-cornered blocks (fig. 2.1).

Figure 2.1. Intersecting joints in bedrock (Courtesy of G. K. Gilbert, United States Geological Survey)

Cracks and fissures greatly increase the exposed surface of the rock. As shown in figure 2.2, slicing a cube in half in each of its three planes produces eight smaller cubes, each with one-fourth the surface of the original cube. As a result, the surface area exposed to the effects of climate is doubled.

Water is an important factor in physical weathering because it expands when it freezes. Deep in the cracks in a rock mass, ice exerts a powerful force that breaks the rock apart in the phenomenon known as *frost wedging* (fig. 2.3). Repeated freeze-thaw cycles can quickly break up any rock that starts out with even the tiniest cracks. By breaking large rock masses into smaller ones, it further increases the surface area that can be attacked by other weathering processes.

Water is also an efficient solvent. The removal of minerals by solution is called *leaching*. Limestone is notably subject to leaching by groundwater. The more acidic the water, the more readily it dissolves calcium carbonate. Great caves such as Carlsbad Caverns are evidence of water's effect on limestone (fig. 2.4).

Most rocks are a mixture of minerals, some of which are more soluble than others. When some of the minerals in a rock have been leached out, the less soluble minerals

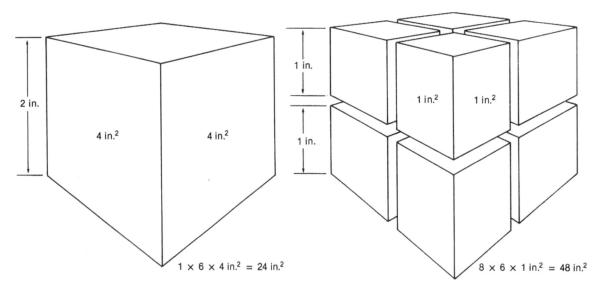

Figure 2.2. Increase in surface area caused by fissuring

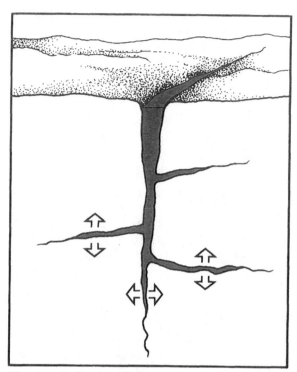

Figure 2.3. Frost wedging

left behind are exposed to other weathering forces such as frost wedging. A well-cemented sandstone can be converted back to loose sand by the leaching of its cementing materials.

Other mechanical weathering agents include plant roots and burrowing organisms, glacial abrasion, lightning, and the impacts of rockfalls and of wind- or water-borne particles.

Chemical weathering. Chemical weathering involves three principal processes: oxidation, hydrolysis, and carbonation. *Oxidation* is the chemical combination of rock minerals with oxygen to form oxides: Just as a sheet of iron rusts and crumbles into a pile of iron oxide, rocks containing iron minerals break down readily as the iron is oxidized.

Hydrolysis and carbonation occur in the presence of water. *Hydrolysis* is a chemical reaction between water and a mineral that breaks down the mineral. *Carbonation* is a reaction with carbon dioxide (in solution as carbonic acid) to form carbonate minerals.

Figure 2.4. Carlsbad Caverns (Courtesy of W. T. Lee, United States Geological Survey)

Both carbonation and hydrolysis are involved in the chemical weathering of orthoclase, a mineral present in some types of granite, as follows:

$$2KAlSi_3O_8 \; + \; H_2CO_3 \; + \; H_2O$$
(orthoclase)   (carbonic   (water)
acid)

$$\longrightarrow Al_2Si_2O_5(OH)_4 \; + \; K_2CO_3 \; + \; 4SiO_2$$
(kaolinite)   (potassium   (silica)
carbonate)

This reaction takes place on an exposed granite surface. Kaolinite, a type of clay, expands when hydrated; soluble potassium carbonate is carried away by water. A similar reaction affects biotite, another mineral in granite. The substitution of clay for these minerals weakens the rock and allows the more resistant grains, such as quartz, to be carried away by erosion. The rock thus breaks down into loose sand composed largely of quartz grains. Less soluble silica may crystallize between the grains of these and other sediments, cementing them together to form sedimentary rock.

Pyroclastics

Pyroclastic particles are commonly called "volcanic ash." Certain types of thick, viscous magma contain much water. When pressure is released, as it was by a great avalanche on the north face of Mount St. Helens on May 18, 1980, the water suddenly flashes into steam, blasting the molten mass into tiny splinters of solidifying glass (fig. 2.5). A fiery cloud of incandescent rock particles, steam, and other gases can flow downhill at speeds up to 200 mph, bulldozing and incinerating everything in its path. Once this momentum is spent, the hot particles come to rest in thick blankets of cooling cinders. Residual temperatures of several hundred degrees may unite the still-soft fragments into *welded tuff*. Thick deposits of this light, porous rock are often found near extinct volcanoes such as the Jemez Caldera in New Mexico (fig. 2.6).

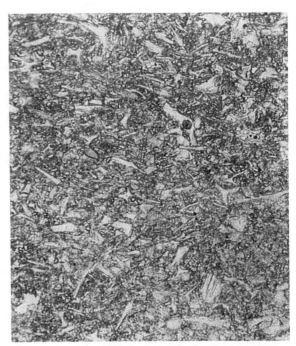

Figure 2.5. Pyroclastic particles, enlarged (Courtesy of W. J. Mapel, United States Geological Survey)

Figure 2.6. Welded tuff near Los Alamos, New Mexico (Photo by J. Morris)

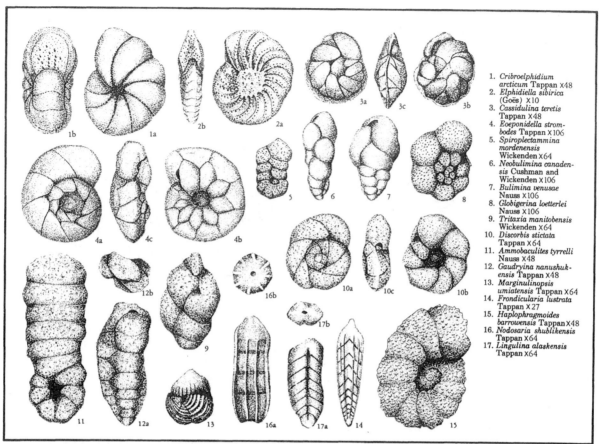

1. *Cribroelphidium arcticum* Tappan ×48
2. *Elphidiella sibirica* (Goës) ×10
3. *Cassidulina teretis* Tappan ×48
4. *Eoeponidella strombodes* Tappan ×106
5. *Spiroplectammina mordenensis* Wickenden ×64
6. *Neobulimina canadensis* Cushman and Wickenden ×106
7. *Bulimina venusae* Nauss ×106
8. *Globigerina loetterlei* Nauss ×106
9. *Tritaxia manitobensis* Wickenden ×64
10. *Discorbis stictata* Tappan ×64
11. *Ammobaculites tyrrelli* Nauss ×48
12. *Gaudryina nanushukensis* Tappan ×48
13. *Marginulinopsis umiatensis* Tappan ×64
14. *Frondicularia lustrata* Tappan ×27
15. *Haplophragmoides barrowensis* Tappan ×48
16. *Nodosaria shublikensis* Tappan ×64
17. *Lingulina alaskensis* Tappan ×64

Figure 2.7. Foraminifera (Drawings by Elinor Stromberg, courtesy of United States Geological Survey)

Nonclastics

Particles of the type that make up *nonclastic* sedimentary rocks are produced differently than the above (clastic) particles. A common nonclastic sedimentary rock is limestone, whose chief component is calcium carbonate (*calcite*). Limestone is formed on the ocean floor. The skeletons and shell fragments of tiny marine organisms, such as *foraminifera* (fig. 2.7), accumulate along with other particles in beds of calcite-rich mud. Other calcareous organisms, such as corals and shellfish, anchor themselves in place and accumulate atop older colonies to form a *reef*. Some limestone is composed of sandlike particles called *oolites*, which are formed when layers of calcite accumulate on smaller particles like ice on a hailstone (fig. 2.8).

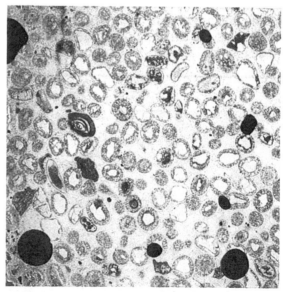

Figure 2.8. Oolitic limestone (Courtesy of P. L. Applin, United States Geological Survey)

Seawater contains dissolved carbon dioxide from the atmosphere. The higher its carbon dioxide content, the more calcite it can hold in solution. Warm seawater can hold less carbon dioxide, and therefore less calcite, than cool water. Particles of calcite that would dissolve before reaching the bottom in colder water will accumulate to great depths in warm water, forming *carbonate muds* that may eventually become limestone.

SEDIMENTARY TRANSPORT

Gravity causes sediments to move from high places to low. Sometimes gravity is the only agent involved, as in a landslide. More often, however, gravity works through another medium—water, wind, or ice—to transport sedimentary particles from one location to another.

The flow of a river moves the sediments in its bed downstream, but it also undercuts its banks (fig. 2.9). When material is continually removed from the lower part of a slope, the slope eventually becomes too steep to support its own weight. It collapses, either gradually or suddenly, into the streambed and is carried away. Thus, although the stream acts directly only on the sediments in its bed, the valley becomes wider at the top because of undercutting.

The inexorable pull of gravity ensures that sediments are carried downhill—ultimately to sea level. However, tectonic forces often

Figure 2.10. Avalanche, Madison Canyon, Montana (Courtesy of William C. Bradley)

raise lowlands high above sea level, ensuring a continuing supply of raw material for producing sediments. Thus any type of rock can become source material for future sedimentary rocks by being uplifted and exposed to weathering and erosion.

Mass Movement

In high mountains, severe weathering and the instability of steep slopes often result in the sudden movement of large masses of rock and sediment. A large block of bedrock may separate along deep fractures or bedding planes, causing a *rockslide* or *avalanche* (fig. 2.10). In seconds, a single rock mass weighing millions of tons can come crashing

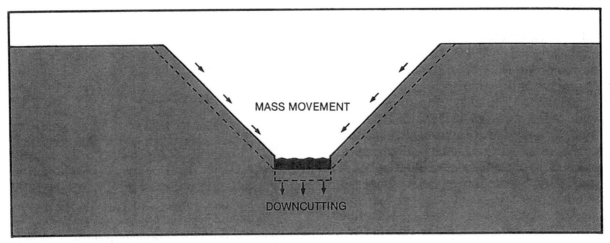

Figure 2.9. Stream downcutting vs. mass wasting

Figure 2.11. Talus (Courtesy of H. W. Malde, United States Geological Survey)

down, shattering itself and everything it hits into boulders and dust.

Although spectacular, rockslides play only a minor part in the mass movement of weathered material. More commonly, frost wedging breaks off rocks and boulders that accumulate in steep *talus slopes* (fig. 2.11). When fresh material is added to a talus slope, it tumbles, slides, and creeps downslope until the maximum stable angle — the *angle of repose* — is reestablished.

Figure 2.12. Slump (Courtesy of J. R. Stacy, United States Geological Survey)

Even though gravity is the principal agent in mass movement, water plays a key part. Besides the destructive effects of repeated freezing and thawing, water adds mobility to weathered rock debris, making it respond more readily to the pull of gravity. Excess water makes clay especially unstable. Under prolonged soaking by rainfall, a clay hillside may suddenly *slump* (fig. 2.12).

Stream Transport

Most sediments, even those involved in mass movement, are at some time transported by flowing water. The distance a sedimentary particle is carried depends both upon the size, shape, and density of the particle and upon the available stream energy.

The steeper a stream, the more gravitational energy it has — and the faster it flows. If the water flows smoothly, like a formation of marching soldiers, the flow is said to be *laminar.* However, if this orderly, parallel movement is disrupted, the flow becomes *turbulent,* tumbling and swirling chaotically. Such a change can be seen where a quiet stretch of river enters a section of shallow, uneven streambed and breaks into turbulent rapids.

It is this turbulent type of flow that makes available most of the energy a stream uses in transporting solid particles. Some of the swirling is directed upward with enough energy to pick up and carry rock particles that are heavier than the water — the same way a high wind can pluck a sheet of tin off a roof. In general, the turbulence of a stream increases with its velocity and with the narrowness and roughness of its channel. Steep, narrow mountain streams have a greater proportion of turbulent energy than sluggish lowland streams, so they are noisier.

Some sedimentary particles are more easily picked up and carried than others. Density is important, of course, but less so than size and shape. All else being equal, a large particle will settle out of still water faster than a small particle. The frictional force that resists the particle's fall (and supports it

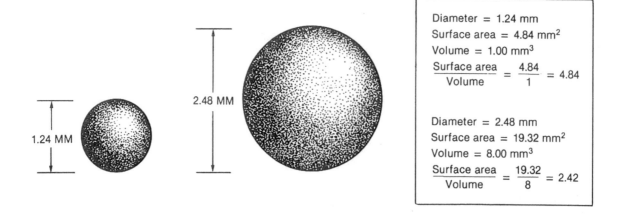

1.24 MM

2.48 MM

Diameter = 1.24 mm
Surface area = 4.84 mm²
Volume = 1.00 mm³
$$\frac{\text{Surface area}}{\text{Volume}} = \frac{4.84}{1} = 4.84$$

Diameter = 2.48 mm
Surface area = 19.32 mm²
Volume = 8.00 mm³
$$\frac{\text{Surface area}}{\text{Volume}} = \frac{19.32}{8} = 2.42$$

Figure 2.13. Surface/volume ratio

in moving water) acts only on the surface of the particle; the smaller the particle, the greater its ratio of surface area to volume and weight (fig. 2.13). The larger particle, with more weight per unit surface, needs more force per unit surface to move it.

The shape of a particle also affects ease of transport. A perfect sphere has more volume and weight per unit surface than any other shape. The less spherical a particle, the more easily a stream can carry it. A flat pebble of quartz will settle like a leaf; a spherical quartz pebble of equal weight will sink more quickly. The difference in shape is like the difference between an open umbrella and a closed one. Opening an umbrella in a high wind does not change its weight, but does make it more likely to be blown away.

For any given shape and size, a denser particle will settle out faster than a less dense particle because its weight-to-surface ratio is higher. For this reason, small particles of gold (specific gravity 16 to 19) collect on the bottom of a streambed, while silt (sp gr 2.5 to 3.0) is swept downstream.

A mixture of sand, silt, and clay in a bucket of water shows how adding flow energy to water affects different types of particles. With the water at rest, the sediment stays on the bottom and the water remains clear. When the water is stirred

vigorously, however, it turns muddy. The added energy is used by the water to pick up and carry the smaller silt and flat, platelike clay particles. The sand grains move mostly by bouncing and rolling along the bottom. As the swirling water slows, the sand comes to rest first. The silt then settles out on top of the sand. The clay tends to remain in suspension long after all motion has stopped; it may take several hours for the water to clear. Even though the flow energy has dissipated, the water resists the downward settling of the clay particles.

The energy added by stirring the bucket of water is like the flow energy imparted to a stream by the force of gravity. In a typical stream, gravel and coarse sand are rolled and bounced along the bottom by the kinetic energy of the flowing water. This material is called the *bed load* (fig. 2.14). Finer sand, silt, and clay are carried well off the bottom by the turbulence of the water; this part is called the *suspended load*. Other products of weathering are carried along in solution – the *dissolved load*.

The energy available for transporting sediments varies over both time and space. Most streams have an annual cycle and carry most of their sedimentary load during the wettest part of the year. Over geologic time, changes in climate may increase or decrease the average flow and therefore the

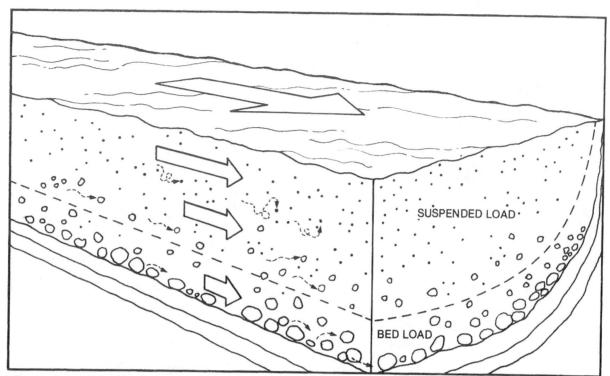

Figure 2.14. Distribution of stream sediment load

total amount of sediment. The velocity of a stream is usually greatest near the center of its cross section; turbulence is greatest along the banks and bed.

Geologic uplifting increases total energy. A steepened gradient increases the stream's erosive power until it has cut deeply enough to regain equilibrium with its load. Many examples of accelerated stream downcutting can be seen in recently uplifted mountains (fig. 2.15).

A *graded stream* is one that is stable, or in balance with its average load. It is just steep enough to carry out of its basin the amount of sediment brought in during an average flow year. Excess sediment tends to be deposited near the head of slow-flowing stream segments, steepening the gradient. Increased erosion then carries away the excess until the stream returns to its original gradient.

Other Transporting Agents

Wind and glaciers are two other agents of transport; compared to flowing water,

Figure 2.15. Accelerated downcutting caused by regional uplift (Courtesy of William C. Bradley)

Figure 2.16. Boulder carved by wind-borne particles (Courtesy of W. H. Bradley, United States Geological Survey)

however, they move only minor amounts of sediment. High winds can carry clay, silt, and sand much as a river does – as suspended load and bed load. In arid climates, wind may even act as the primary weathering and transporting agent, carving exposed rock into fantastic shapes by abrading it with airborne particles (fig. 2.16). A glacier moves slowly but with great weight, grinding rocks into powder and carrying jumbles of unsorted rock material to its snout (fig. 2.17). Both wind-driven and glacial sediments are often reworked and redeposited by flowing water.

DEPOSITIONAL ENVIRONMENTS

When sedimentary particles arriving at a location outnumber those being carried away, the location can be thought of as a *depositional environment*. Depositional environments are of many different types. Wind, waves, temperatures, flow patterns, seasonal variations, and biotic communities are a few of the many factors that influence the character of the sedimentary deposits in a given environment.

Continental Environments

Because most areas above sea level are subject to erosion, depositional conditions tend to be more localized onshore than offshore. They are also more affected by rapid

Figure 2.17. Glacial deposition (Courtesy of T. L. Pewe, United States Geological Survey)

changes in weather. At certain places and times a river deposits more sediment in its valley than it carries off. However, the sediments deposited during months of slack water may be carried downstream in a single day of heavy runoff.

Fluvial deposits. Sediments deposited by flowing water are called *fluvial* deposits. Local variations in flow determine where particular types of sediments accumulate (fig. 2.18). Stream velocity is greatest on the outside of a bend, where the stream undercuts the bank and increases its sedimentary

load. On the inside of the bend, where the water slows and eddies, the stream has less energy. Here is left much of the suspended sediment—first the sand, which takes the most energy to move; then higher on the sloping bar, silt; and farthest from the swift main current, clay. Bed load materials such as cobbles and gravel collect in deeper water near the base of the sand.

A river in flood uses much of its increased energy to augment its suspended load. Adding silt and clay from its channel to the materials coming from upstream, the river

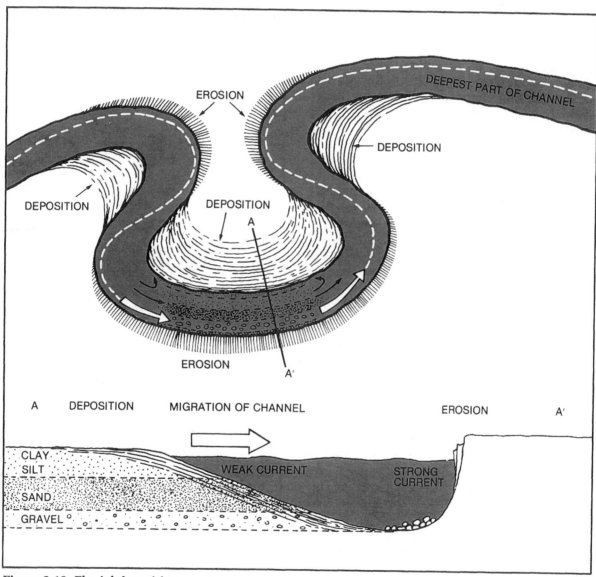

Figure 2.18. Fluvial deposition

Figure 2.19. River in flood (Courtesy of I. D. Yost, United States Geological Survey)

spreads out over its floodplain (fig. 2.19). Here friction absorbs much of the flow energy. Silt and clay settle out, raising the level of the floodplain (and enriching the soil for plant life). Some of the heavier sediments accumulate atop the banks nearest the river, forming natural *levees* that help contain the river at lower flow stages.

Sandbars and other stream deposits overlap one another in characteristic ways. A cross section of such overlapping deposits reveals their characteristic lens shape — thick in the middle and tapering toward either edge (fig. 2.20). Bar deposits are long and curved in the direction parallel to the stream but narrow and lens-shaped in section across the stream. Evidence of flow

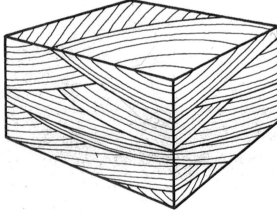

Figure 2.20. Overlapping lenses characteristic of stream sediment deposits

direction and volume is preserved in the form of *ripple marks, scour marks,* and other structures.

A meandering river creates an even more complex system of overlapping and under-cut deposits. *Meandering* occurs when a river with excess energy and a flat flood-plain erodes one of its banks more than the other and begins shifting in a gentle curve toward that bank (fig. 2.21A). Since a curved line connecting two points is longer than a

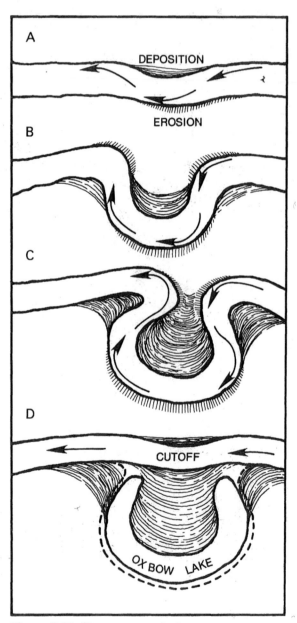

Figure 2.21. Stream meandering

straight line, a meander reduces the river's gradient: the river travels farther to descend 1 foot in elevation. As the curve deepens, the opposite bank is eroded at its beginning and end (B). With continued erosion and deposition, the meander takes the form of a narrow-necked loop (C) that the river may eventually erode through to form a *cutoff* (D). In the *oxbow lake* thus formed in the abandoned loop, sediments and life forms differ from those of the river.

Lacustrine environments. The still waters of a lake absorb all the flow energy of inflowing rivers, causing the rivers to deposit their sediments near their entry points. In addition to the inflow of mineral sediments, dissolved nutrients flow in and feed the growth of a biotic community of plants and animals. The remains of these organisms accumulate in the sediments of the lake bottom rather than being flushed downstream as in a river. Eventually the lake fills with sediments and ceases to exist, leaving behind a deposit from which fossil fuels such as coal or oil may be born.

Desert environments. In arid climates, infrequent downpours and flash floods leave sheets of gravel and sand in large, sloping deposits at the mouths of canyons. This type of deposit is called an *alluvial fan* (fig. 2.22). Such *fanglomerates* (some of which are termed *molasse*) are thin, overlap-ping, poorly sorted sheets of angular gravels, boulders, and mud. A line of fans may eventually coalesce into an *apron* that grows broader and higher as the slopes above are eroded.

In enclosed desert basins, scarce runoff may create intermittent playa lakes, also known as *sebkhas* (or *sabkhas*). Coarser sediments are deposited around the margins of a sebkha in alluvial fans and aprons; silt and clay are carried into the central parts, where they settle out more slowly. When the water evaporates, dissolved salts crystallize out to form thin crusts of *halite* (rock salt), *gypsum* (hydrous calcium sulfate), or other *evaporites*. A sebkha thus develops a characteristic pattern of alternating thin beds of mud and evaporites.

Glacial deposits. Sediments deposited by moving ice sheets are much rarer than other types, principally because deposits created by geologically infrequent ice ages are subject to erosion and reworking by other agents. Retreating glaciers and ice sheets leave behind accumulations of unsorted sediments called *till*. Glaciers grind bedrock into flour and carry along great boulders that a river would simply flow around. Glacial till is thus recognizable by its chaotic jumble of mud, gravel, and large rocks. When a glacier retreats, meltwater usually reworks and redistributes the till. Glacial till and outwash sediments from the last ice advance cover much of the northeastern United States. Buried sediments from older glacial periods can be found worldwide, often in locations now too far from the poles to have been affected by more recent ice ages.

Aeolian deposits. Sediments deposited by wind (*aeolian* deposits) range from *loess* (thick beds of silt carried by winds from the outwash plains of glaciers) to sand dunes. Large desert dunes that migrate downwind because of prevailing winds develop a pattern of cross-bedding that may be preserved beneath younger sediments. Examples of cross-bedded dune structure can be seen in canyons of the southwestern United States (fig. 1.6).

Figure 2.22. Alluvial fans (Courtesy of J. R. Balsley, United States Geological Survey)

Transitional Environments

Most of the sediment produced by the weathering of rock is carried by flowing water to the sea. Here the fresh water of the river quickly becomes dispersed in the much larger volume of salt water—as does flow energy. Sediments are deposited throughout this *transition* zone, where the river gradually gives up its energy; they are sorted by size, shape, and density according to the energy distribution of their environment, as they were in upstream depositional environments.

Marine deltas. The mouths of rivers fall into two general categories. A river with a low sediment load may reach the sea in an *estuary*, where the effects of current, waves, and tides keeps sediment from accumulating. A heavily laden river, however, usually creates a *marine delta*, a seaward extension of land at or near sea level caused by the accumulation of sediments at the river's mouth (fig. 2.23). A delta is composed mostly of sediments brought down by the river, but mixed with them are fine wave-borne sediments brought in by currents and tides.

A growing (*prograding*) delta is a complex structure of interbedded sediments with three more or less distinct zones of deposition. The zone nearest to the shore is occupied by *topset* beds, complexes of heavier, coarser particles. Depending upon flow energy and sediment load, these are typically overlapping sand and gravel bodies forming seaward extensions of the stream's natural channel and levees. They often show ripple marks, erosion surfaces, and other evidence of flow. Shallow bays between diverging (*distributary*) channels contain finer sediments from both the river and the sea, and often shelter abundant plant and animal life.

Some of the finer sediments carried beyond the topset beds settle out in *foreset* beds on the steep seaward face of the delta. In the upper foreset zone, where wave action suspends the finer sediments, the foreset beds may consist of clean sand; below the wave base, muddy sand and silt; and near the toe of the delta, thinner beds of silt and mud. As the delta grows seaward, the foreset beds are overlain by extensions of the topset beds.

Bottomset beds are tapering layers of silt and clay extending seaward from the face of the delta. Beyond reach of river and waves, these sediments accumulate slowly and sometimes support considerable seafloor life.

The delta described above is an idealized model that is rarely found in nature, where conditions vary widely. Distinct topset, foreset, and bottomset beds are more often seen in lacustrine deltas in sheltered continental environments.

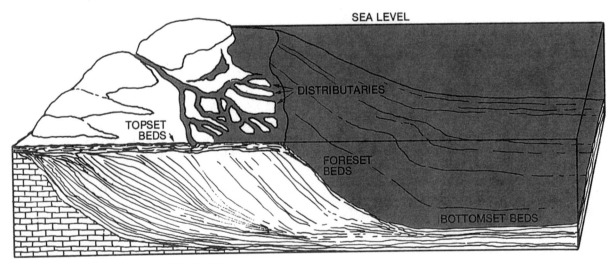

Figure 2.23. Marine delta (idealized)

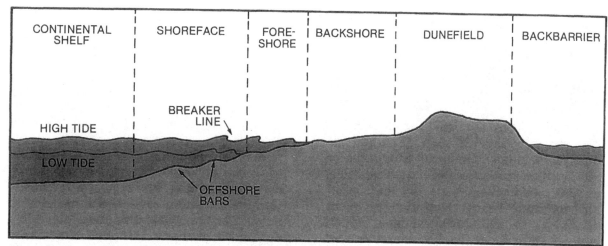

Figure 2.24. Profile of beach, showing energy zones

Beaches. The transition zone also includes *beaches,* coastal depositional environments unrelated to the mouths of rivers. Sand and finer sediments are redistributed by *longshore* currents (the movement of seawater parallel to the coast). The growth and shrinkage of nondelta coastal deposits are related to the energy of longshore currents, waves, and tides.

A cross section of a typical beach shows how the sediments brought in by longshore currents are sorted and redistributed by wave and tidal energy (fig. 2.24). Differences in energy levels divide the beach profile into the *shoreface* (the wave action zone up to the low-tide mark), the *foreshore* (between low- and high-tide levels), the *backshore* (from high tide to storm-flood level), and the *dunefield* (above storm-flood level). Because these zones are continually affected by wind and waves, their sediments are usually clean, well-sorted mineral grains and shell fragments. A cross section of a beach dune may show the same excellent sorting and cross-bedding found in desert dunes. Coarser gravels are deposited in deeper water, and finer sediments are carried seaward by wave action or inland by wind.

As the shoreface slopes upward toward the land, ocean waves begin to "feel" the bottom—that is, friction retards the movement of water near the bottom of a wave.

The top of the wave goes ahead of the bottom, causing the wave to topple, or "break." While at the seashore you may have noticed a line of breakers some distance offshore. A breaker line marks an underwater bar where the concentration of wave and backwash energy causes an accumulation of coarser sediments, such as gravel and seashells. One or more bars may form on the shoreface, depending upon the energy of incoming waves and the availability of sediments.

Another high-energy zone occurs where the incoming wave laps up onto the foreshore. This action sorts the beach sediments, washing the sand clean of finer sediments and leaving particles of similar size together in distinct zones. Sediments fine enough to be suspended in the turbulent waves are carried offshore and deposited in quiet water beyond the breaker line.

If the beach is on an offshore barrier (spit or island), the transition zone may also include a shallow lagoon where sediments accumulate in a *backbarrier complex.* Such depositional environments are highly variable and may include tidal channels, salt marshes, shell reefs, and mangrove swamps, among other features.

Deposition in the backshore zone is intermittent. Coarser sediments may be deposited here when storm waves surge over the highest beach bars and run back into the sea in shallow channels parallel to the shore. Dry sand in the backshore zone may migrate inland in desertlike dunes driven by onshore winds.

If the lagoon behind the barrier is relatively isolated from the sea, the backbarrier complex may include such disparate features as sebkhas, thin-bedded deposits of salt or other evaporites formed where intermittent, landlocked pools dry out; peat, formed from heavy deposition of plant debris (as in mangrove swamps), which may, if buried deeply, become coal; and organic muds rich in carbonates and other skeletal debris. On the other hand, strong currents behind the barrier island or spit may flush out these deposits, leaving sediments in configurations much like those in stream deposits.

Marine Environments

Marine depositional environments are those seaward of the beach — that is, beyond the zone normally affected by wave action or fluvial deposition. Associated with low-energy environments, marine sediments are mostly finer than those in the transition zone.

Shelf environments. A typical *continental shelf* is an underwater plain sloping 10 feet per horizontal mile seaward. It extends from the transition zone (beach or delta) to the edge of the continental mass (fig. 2.25). Most of the energy in this zone comes from shallow ocean currents and tides. Sediments are mostly silt and clay mixed with fecal pellets, skeletal debris, and shell fragments, with local deposits of coarser sediments brought in by infrequent tidal and flood currents. In warm tropical oceans, the clay of

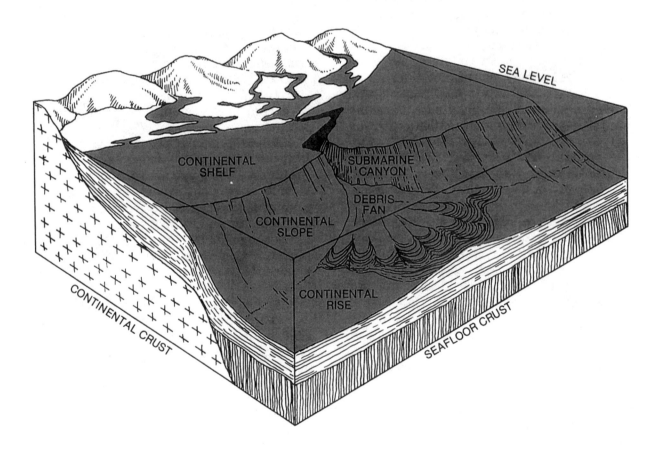

Figure 2.25. Continental shelf, slope, and rise (simplified)

the shelf often grades into carbonate mud made up of undissolved shell and skeletal debris.

An *epeiric sea* is a broad, shallow arm of the ocean extending well inland upon the continental platform. An epeiric sea accumulates sediments from the land and organic matter from marine life, like the continental shelf, but is isolated from oceanic currents and tides. The only known modern examples, Hudson Bay and the Baltic Sea, are both cold epeiric seas. In the geologic past, however, great warm, shallow seas covered most of North America. Sediments accumulated to thicknesses of several miles as their weight gradually depressed the continental bedrock beneath. Some of these sediments were later thrust high above sea level, where they are visible today in the Rocky Mountains and the Grand Canyon.

A *reef* is a wave-resistant deposit consisting of the calcareous remnants of marine organisms in or near the locations where they grew. A coral reef is a familiar example; however, the term *reef* is also applied to algal and oyster mounds. A reef dissipates wave energy like an offshore sandbar. Calcareous debris and organic matter may accumulate in the relatively quiet water between the reef and the shore. The deeper water seaward serves as a basin for detritus broken off the reef by wave action.

In the northern part of the Permian Basin of West Texas can be found an outstanding example of a buried reef: the Horseshoe Atoll (fig. 2.26). Roughly circular and about 80 miles across, this formation consists of

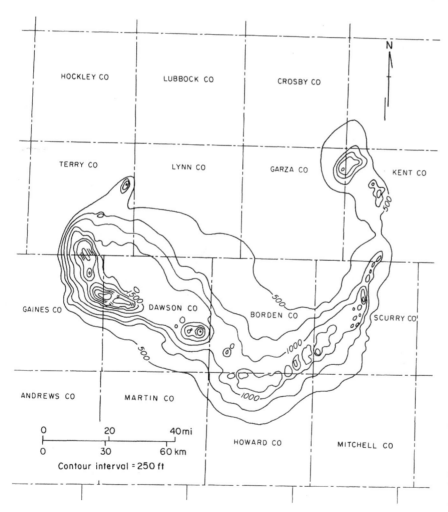

Figure 2.26. Structure contour map of Horseshoe Atoll (From *Atlas of Major Texas Oil Reservoirs* by W. E. Galloway, et al., 1983. Courtesy of Bureau of Economic Geology, The University of Texas at Austin)

Figure 2.27. Model of Permian Basin reef (Courtesy of Permian Basin Petroleum Museum. Photo by T. Gregston)

up to 3,000 feet of fossiliferous limestone. It was laid down during Pennsylvanian and early Permian times atop a limestone platform in a warm epeiric sea. (Figure 2.27 shows a model of a typical Permian Basin reef with some of its associated organisms.)

Late in the Permian era the reef was buried beneath more than a mile of clay, sand, and evaporites. El Capitan, in the Guadalupe Mountains of West Texas, is an

example of such a reef that has been uplifted and exposed by erosion (fig. 2.28).

Outer continental environments. Beginning at the edge of the continental shelf, typically in 600 to 800 feet of water (although sometimes much more or much less), is the *continental slope*. Unlike the smooth continental shelf, the continental slope is a zone of steep, variable topography forming a transition from the shelf edge to

Figure 2.28. El Capitan, a former reef in West Texas (Courtesy of P. P. King, United States Geological Survey)

the continental rise (fig. 2.25). Its average slope is around 4°, a drop of 350 feet per horizontal mile. It is also dissected, near the mouths of rivers, by rugged submarine canyons. Sediments deposited on the slope, like the foreset beds of deltas and dunes, are relatively unstable and tend to migrate toward its base.

The transition zone between the continental slope and the oceanic abyss is the *continental rise*. A major mover of sediments on the continental rise is the *turbidity current*—a dense mass of sediment-laden water that flows down the continental slope, typically through a submarine canyon. Fed by a flooding river, underwater debris slide, or other source, a turbidity current is a sort of underwater flash flood carrying clay, silt, and gravel rapidly downslope in a narrow tongue. Its energy is quickly dissipated by friction upon the more gradual slope of the continental rise, where its sediments settle out, largest particles first, to form a *turbidite*—a thick, graded bed topped by clay (fig. 2.29).

A succession of turbidites forms a *flysch* deposit. Submarine debris fans along the base of the continental slope coalesce into an apron that is interbedded with a continuous fallout of fine sediments. Extensive flysch deposits in a geologic column tell the geologist that the environment was once the outer margin of a continent.

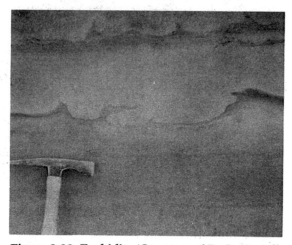

Figure 2.29. Turbidite (Courtesy of D. G. Howell, United States Geological Survey)

SEDIMENTARY FACIES

The sediments deposited in a continuous sheet across a broad area differ from one location to another. At the edge of an epeiric sea, for instance, the layer being deposited at any given time may grade continuously from beach sand, through a zone of silt and clay, and into a lime mud and reef complex some distance offshore. Salt may accumulate on the bottom in restricted areas. These deposits will form, respectively, sandstone, siltstone, shale, limestone, and halite. The transition from one rock type to another within the layer is continuous and gradual; a geologist tracing the layer from one area to another might find it hard to

determine where limestone ends and shale begins (fig. 2.30A). These changes in rock character are called *facies changes*. Facies changes often occur within a single *formation*, which is a lithologically distinctive rock body, with an upper and a lower boundary, that is large enough to be mapped.

If sea level rises relative to the land, each depositional environment will follow the retreating shoreline inland. Each type of rock thus formed will grow laterally toward the coast (fig. 2.30B). Instead of a vertical column, the limestone will form a continuous, near-horizontal sheet stretching shoreward. This layer might be treated, and even named, as a distinct formation. In reality, however, it is merely one facies of a

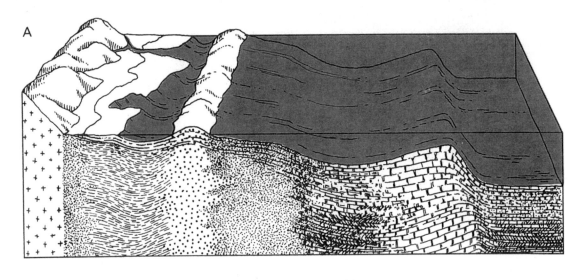

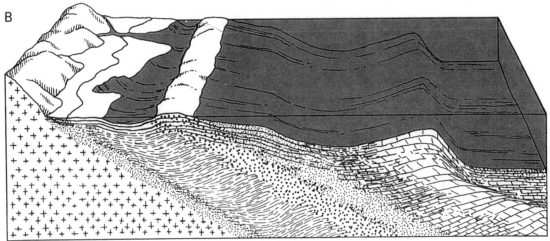

Figure 2.30. Facies

larger unit and was formed over a long period of time concurrently with the facies above and below it. The geologist must distinguish between a *rock stratigraphic unit,* such as the limestone in this example, and a *time stratigraphic unit,* the concurrently deposited layer with all its lateral facies variations.

LITHIFICATION

Rivers flow in only one direction—down to the sea. Unless the land subsides or sea level rises, sediments deposited in fluvial environments above sea level are eventually carried away by the unrelenting erosive force of the river. However, sediments that reach the sea eventually come to rest in a relatively stable environment, beyond the influence of most wind, weather, and current, and become buried beneath other sediments. For this reason most sedimentary rocks originate in marine environments.

Once deposited, sediments are not necessarily *lithified* (transformed into stone). In order for lithification to occur, unconsolidated sediments must remain beyond the reach of erosion, and they must be compacted and cemented. The physical and chemical changes that sedimentary deposits undergo during and after lithification are known collectively as *diagenesis.* (Diagenesis does not, however, include the more radical changes associated with the heat and pressure of metamorphosis.)

Compaction

If the accumulation of sediments continues over a long period, as it usually does in the ocean, great thicknesses of material may be placed on top of the original layers. Burial beneath thousands of feet of other sediments is what begins to turn sediments into rock. The weight of overlying layers squeezes particles together into the tightest arrangement possible; under extreme conditions, it may crush some or all of the grains. As the water that originally filled the pore spaces between particles is squeezed out,

the volume occupied by the sediments is reduced. This process is called *compaction.*

Different clastic materials pack differently. A dune sand composed entirely of rounded grains of similar size may be compacted very little (fig. 2.31A). In a poorly sorted sand, however, smaller particles can be rearranged to fill the spaces between the larger grains, thus reducing the pore space. Platelike clay particles, however, become aligned with one another when compacted, fitting together like bricks with very little space between (fig. 2.31B). Some clays become compacted to less than half their original volume under the pressure of deep burial.

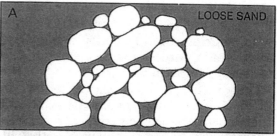
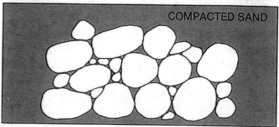
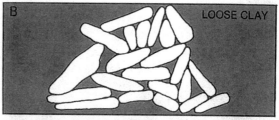
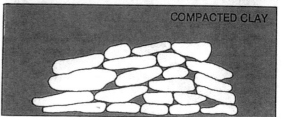

Figure 2.31. Compaction of sand and clay

Cementation

As compaction brings individual particles into closer contact, the process of lithification is completed by cementation. Minerals in solution – mainly calcite ($CaCO_3$), the basic constituent of limestone – crystallize out of solution to coat the grains (fig. 2.32). Other common cementing agents include silica (SiO_2), which is less soluble and therefore less abundant in groundwater; and iron oxide (Fe_2O_3), which colors the rock yellow or red. These coatings grow together and may eventually fill the pore spaces. Clean sand may be transformed into limy sandstone, losing most of its original pore space. Because of its low permeability, compacted clay can take on only a small amount of cement in the process of becoming shale; but wet clay is cohesive to begin with, sticking together by its own internal electrostatic forces, and so requires less cementing to become rock.

Cementation is the crystallization or precipitation of soluble minerals in the pore spaces between clastic particles. A defining characteristic of clastic texture is that individual particles touch each other at various points without conforming to the shape of other particles. In crystalline texture the grain structure originates in the rock itself as the crystals of various minerals grow in close contact with each other (fig. 1.24). Igneous and metamorphic rocks are crystalline and therefore have little or no pore space.

The texture of cemented sedimentary rock is part clastic and part crystalline. Clastic sedimentary rock can range from a lightly cemented porous sandstone to a well-cemented limy sandstone. Some sandstone contains so much calcite that it may be considered a sandy limestone rather than a limy sandstone.

Carbonate sedimentary rocks display an even wider range of textures because calcite particles can be created in a variety of ways. They can be the products of the mechanical breakdown of corals, shells, and skeletons, or they can form as oolites, the egg-shaped rounded grains created by the deposition of layers of calcite on other particles. As they collect in sedimentary basins, these particles can become cemented together in the same way as other grains – by crystallized calcite. The result is a limestone consisting of calcite fragments in crystallized calcite. Depending upon the admixture of other particles, limestone can grade through sandy, silty, or shaly limestone into limy sandstone, siltstone, or shale.

The deposition of sediments occurs in the earth's *biosphere* – the thin zone of air, water, and soil where all terrestrial life exists. For this reason, sedimentary rock often contains fossils – animal or plant parts, entire organisms, or such evidence of their former presence as tracks or burrows. Limestone, in particular, is formed partly from the remains of calcareous organisms and often contains an abundance of their shells (fig. 2.33).

Once an accumulation of sediment has become compacted and cemented into the durable form of true rock, it can undergo further changes in its chemical and physical environment that alter its structure and composition. These diagenetic alterations can involve the leaching of soluble minerals (increasing porosity), the addition of minerals by crystallization (decreasing porosity), or the recrystallization of the

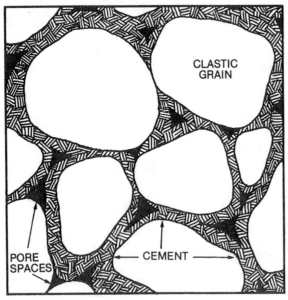

Figure 2.32. Cementation

Figure 2.33. Shelly limestone (Courtesy of E. B. Hardin, United States Geological Survey)

Figure 2.34. Conglomerate (Courtesy of C. C. Albritton, Jr., United States Geological Survey)

minerals present in the rock itself, as in the formation of chert in siliceous shale.

Diagenesis can also involve the chemical replacement of one element or mineral by another. Limestone, for example, is changed to *dolomite* (or more properly, *dolomitic limestone*) by replacement of half or more of its calcium with magnesium. Dolomitization may occur when limestone is saturated with magnesium-rich groundwater, especially at higher temperatures.

CLASSIFICATION OF SEDIMENTARY ROCKS

Geologists often divide sedimentary rocks into three main groups: clastics, evaporites, and carbonates. However, due to the many chemical and physical processes that affect sediments during lithification and diagenesis, most sedimentary rocks have features of more than one of these types.

Clastics

Clastics are rocks such as conglomerates, sandstones, and shales that are composed mostly of fragments of other rocks. The principal distinction among clastics is grain size.

Conglomerates. *Conglomerates* are rocks most of whose volume consists of particles more than 2 mm in diameter (fig. 2.34). They are formed from the sediments found in alluvial fans, debris flows, glacial outwash, and other high-energy depositional environments. Most conglomerates occur in thin, isolated layers; they are not very abundant.

In common usage, the term *conglomerate* is restricted to coarse sedimentary rock with rounded grains; conglomerates made up of sharp, angular fragments are called *breccia* (fig. 2.35). The rounding of fragments usually indicates transport by flowing water,

5 CM

Figure 2.35. Breccia (Courtesy of W. J. Moore, United States Geological Survey)

Figure 2.36. Conglomerate, showing imbrication (Courtesy of D. A. Lindsey, United States Geological Survey)

which tumbles and polishes the grains and breaks off sharp projections. The rounder a pebble, the farther it has been transported. Stream gravel deposits often show *imbrication*—the arrangement of pebbles in a flat, overlapping pattern like bricks in a

Figure 2.37. Sandstone (Photo by J. Morris)

wall (fig. 2.36). Conglomerates formed from debris slides and alluvial fans have more angular fragments and poorer sorting than those in stream deposits. *Tillite,* a breccia deposited by glaciers, is a chaotic jumble of grains ranging from clay particles to angular boulders; it shows little or no grading or sorting.

Sandstones. *Sandstone* is clastic sedimentary rock more than half of whose grains are $1/16$ mm to 2 mm in diameter (fig. 2.37). About one-fourth of all sedimentary rock is sandstone. The sand particles are of three principal types: quartz (SiO_2); feldspar grains, derived from granite or other rocks; and *lithic* particles, each of which is a mixture of various minerals. Sandstones are of particular interest to petroleum geologists because they are usually more permeable to formation fluids than other rocks and therefore able to accumulate oil and gas. Many of the world's richest petroleum deposits are found in sandstone reservoirs.

Sandstone can remain mostly open and porous or become filled with cement, silt, or clay. Sandstones in which less than 15% of the total volume is silt and clay are called *arenites.* Quartz arenites tend to develop from deposits of clean sand in high-energy environments such as deserts and beaches. Their grains are rounded and well-sorted, indicating much reworking by wind and water. *Arkose* is a feldspathic arenite derived from granite and lithified in small deposits (*granite wash*) near its source.

Graywackes are sandstones that contain more than 15% silt and clay. Their grains tend to be angular and poorly sorted. In some cases, their clay/silt matrix is believed to have been deposited along with the larger grains, as in the heterogeneous mixture of particles found in turbidity currents; in others, clay and silt appear to have been added to the rock by physical or chemical alteration of the original minerals or by infiltration. Most graywackes are thought to be marine in origin.

Shales. An estimated one-half to three-fourths of the world's sedimentary rock is shale—a distinctive, fine-grained, evenly bedded rock composed of silt and clay

Figure 2.38. Shale (Courtesy of W. H. Bradley, United States Geological Survey)

(fig. 2.38). Silt grains are similar to sand grains, but much smaller—$1/256$ mm to $1/16$ mm—and mostly invisible to the unaided eye. Clay particles are quite different in composition and shape; they are microscopic, flat, platelike crystals less than $1/256$ mm across. During compaction and lithification, these plates tend to become aligned in horizontal sheets. As a result, most shale splits along well-developed planes parallel to, but not necessarily coinciding with, bedding planes. One variety, *mudstone,* is an exception; it breaks into chunks or blocks.

Formed from fine sediments that settle out of suspension in still waters, shale and mudstone occur in thick deposits over broad areas, often interbedded with siltstone, sandstone, or limestone. They vary widely in color, with the darker colors often indicating higher proportions of organic materials from the remains of plant and animal life. Petroleum geologists believe organic shales to be the source of most of the world's petroleum and natural gas. Shales also make excellent barriers to the migration of fluids and therefore tend to trap pools of petroleum in adjacent porous rock.

Evaporites

Rocks formed by the precipitation of chemicals from solution are called *evaporites* (fig. 2.39). Deposits of evaporites can form as the result of either the drying up of a body of water, such as a desert playa lake, or continuous evaporation from a confined body of water that is continually replenished by inflowing water. In the latter case, dissolved minerals become supersaturated (so concentrated that they can no longer stay in solution) and precipitate out to form deposits on the bottom. Evaporites usually form in a distinct sequence, the least soluble minerals precipitating first, the most soluble last. A typical deposit would have gypsum ($CaSO_4 \cdot 2H_2O$) and anhydrite ($CaSO_4$) at the bottom, followed by halite or rock salt ($NaCl$), sodium bromide ($NaBr$), and potash (KCl).

Figure 2.39. Outcrop of evaporite (Courtesy of W. W. Mallory, United States Geological Survey)

Evaporites are indicators of former dry climates or enclosed drainage basins. They comprise only a small fraction of all sedimentary rocks but play a significant part in the formation of certain types of petroleum reservoirs—those associated with salt domes.

Carbonates

Most carbonate rocks are organically deposited—that is, formed as a direct result of biological activity. Limestone and dolomite are the most common of these, making up about one-fourth of all sedimentary rocks. Carbonate rocks are important to the petroleum industry; many of the giant reservoirs in the Middle East are in limestone.

Limestone forms in warm, shallow seas. Most seawater is nearly saturated with dissolved calcite ($CaCO_3$); many marine organisms use this calcite to build their shells and skeletons. In shallow tropical oceans, two factors favor the formation of limestone: (1) as water gets warmer, it loses some of its ability to hold dissolved carbon dioxide, and (2) photosynthetic algae and other plants remove CO_2 from the water to produce carbohydrates. This loss of CO_2 reduces the water's ability to hold calcite in solution. Thus the calcareous remains of dead organisms do not dissolve as they would in colder water. Instead, they accumulate in thick layers of lime mud and sand on the bottom, cemented by additional calcite precipitated from solution. If coral is present, the skeletons of large colonies may be engulfed in other calcareous debris.

Great sheets of limestone formed in past epeiric seas (most notably those of the Cretaceous period) now extend across much of the world's land area. Their structures vary from mostly clastic to mostly crystalline and from very porous (with many large openings) to very tight, depending upon where and how they were formed and what has happened since. Most of the limestone has undergone extensive alteration; diagenesis is especially common in limestone because it is more soluble than other sedimentary rocks.

Some other rocks of biochemical origin are *diatomite,* an accumulation of the siliceous (glassy) shells of certain microscopic algae, which sometimes recrystallizes as flinty nodules of *chert; phosphorite,* composed largely of calcium phosphate from bird droppings and vertebrate skeletal remains; *coal,* the mildly to strongly metamorphosed remnants of undecayed plants; and *oil shale,* consisting partly of unoxidized plant and animal remains. Oil shale and other organic shales are primarily clastic rocks, but they contain enough preserved organic matter to be of interest to the petroleum geologist.

STRATIGRAPHY

When sediments settle out of a transporting medium, they form distinct layers, or *strata,* like the layers of snow from a series of winter storms. These strata can most easily be seen where exposed by erosion (fig. 2.40). The thickness of each layer

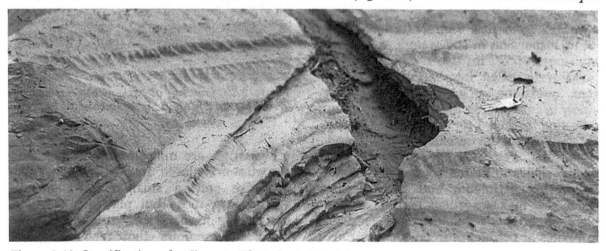

Figure 2.40. Stratification of sediments (Photo by J. Morris)

(*stratum*) depends upon how much material is deposited before deposition stops. A brief shower in the desert leaves a ¹⁄₁₆-inch sheet of silt in a playa lake; the evaporating water, in turn, deposits a thin film of salt atop the silt. Several inches of sand grains, all the same size, accumulate in a large river delta during a spring month; heavy regional flooding gives rise to offshore turbidity currents that leave graded deposits many feet thick on the continental rise.

The major characteristic that sedimentary deposits have in common is the tendency to be formed in strata that are essentially horizontal. When deposited in a valley, for instance, the layers do not form even sheets up the sides of the valley, like snow, but horizontal layers that lap out on the existing slope like water in a lake (fig. 2.41).

By studying the rocks of the crust, the geologist can learn much about what happened in the geologic past. Igneous and

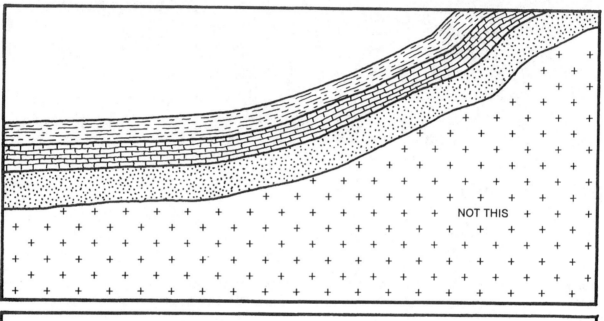

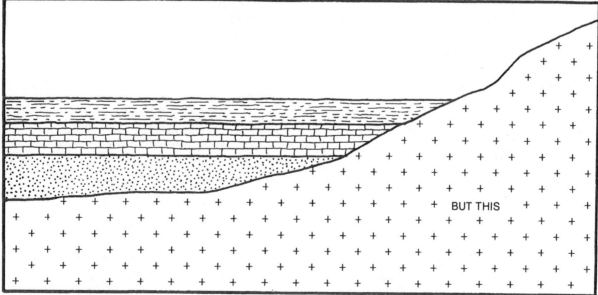

Figure 2.41. Law of original horizontality

Figure 2.42. Limestone strata near Austin, Texas (Photo by T. Gregston)

metamorphic rocks tell him something about major geologic events such as continental rifting or collisions. More detailed studies of individual igneous or metamorphic rocks can indicate whether the rocks were formed on the surface or at depth, how long they took to cool and solidify, whether they were formed from magma beneath the oceans or within the continent, or whether they came from other types of rocks that were altered by great pressure.

The most eloquent story, however, is told by sedimentary rocks (fig. 2.42). Because the materials from which they are formed are deposited on the surface of the earth by forces in the local environment, they can show what was happening to that surface hundreds of millions of years ago. The shape, size, and composition of individual clastic grains provide clues about where they originated and what direction the water, wind, or ice that brought them must have come from. And because all organisms lived on or near the surface, sediments often include evidence of past life. *Fossils,* lithologic traces and remains of life forms, can be found in all kinds of sedimentary

rocks; their types and abundances give evidence about the environments in which the organisms lived and the sediments deposited. Changes in climate, topography, and continental location over the millenia can be mapped – except, of course, where the record has been erased by erosion. The vertical mile of cliffs in the Grand Canyon displays fragments of a geologic history extending from Precambrian through Triassic times – about 2 billion years in all.

Correlation

A formation is customarily named for the location where a distinctive outcrop is first identified. The Morrison Formation, for example, is a variable group of sedimentary rocks found from the Grand Canyon to Wyoming and eastward to Morrison, Colorado, where it was first identified and named. Similarly, the Austin Chalk is a soft, white limestone underlying much of East Texas but named for an outcrop in Austin (fig. 2.43). As we saw earlier, however, it is important to distinguish between rock stratigraphic units and time stratigraphic units. A formation does not necessarily

Figure 2.43. Outcrop of Austin Chalk in Austin, Texas (Photo by T. Gregston)

represent a body of rock that was deposited or formed all at one time. The depositional environment responsible for creating the formation may move gradually from one location to another, like a paving truck laying fresh asphalt over an old highway.

Correlation is the process of relating the rocks at one site with those at another. Often the geologist can find outcrops in only a few places many miles apart—a cliff here, a canyon wall there, rock samples from a well or two (fig. 2.44). Perhaps the cliff has a

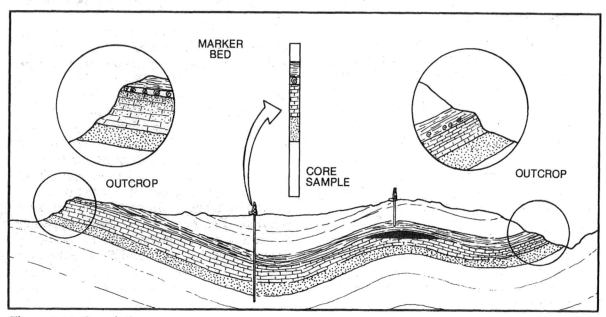

Figure 2.44. Correlation

thick layer of sandy limestone with a layer of shale above and conglomerate below. How does the geologist determine that the shaly limestone in the canyon wall is part of the same formation as the sandy limestone of the cliff? If both are bounded above and below by similar layers of shale and conglomerate, the chances are good that they are different facies of the same formation. If both assemblages contain the same *marker bed*—a formation with a distinctive assemblage of fossils, for example—he can be even more certain of the correlation.

The character and dimensions of rock formations almost always vary from location to location. The thick, massive limestone of a reef thins rapidly seaward into sloping beds of calcite debris and fossils; toward the former shoreline, it may pass into a shaly limestone facies with stringers of silt, then into limy shale. With sufficient information to fill in the gaps, a geologist may see that the shale he finds at a particular place in the geologic column is laterally continuous with (and part of the same time stratigraphic unit as) the massive limestone reef identified in a stratigraphic test miles away. Such information is of great significance in exploring for oil and gas.

Superposition

Reading the geologic record is mostly a matter of applying logic and common sense, although what we take for granted today was by no means easy for geologists or anybody else to see at first. The law of *superposition*, first formulated by Nicolaus Steno in 1669, is an example. It can be stated as follows:

In an undisturbed sequence of sedimentary rock layers, each layer is younger than the one below and older than the one above.

This principle did not become thoroughly accepted until it was well understood that the layered rocks in question were formed from sediments deposited atop other sediments, like books stacked one on top of another. Distinct layers of sedimentary rock

Figure 2.45. Bedding planes (Courtesy of M. N. Bramlette, United States Geological Survey)

can be as thin as a sheet of paper or many tens or even hundreds of feet thick. What usually distinguishes a formation, as the term is commonly used, is a body of rock thick enough to be plotted on a geologic map, with distinctly different rocks above and below.

A *contact* is any sharp or well-marked surface between two different formations. A sedimentary rock formation usually contains many *bedding planes*, more or less distinct surfaces caused by minor changes in sediment type or depositional conditions (fig. 2.45). A bedding plane is essentially horizontal because of the leveling action of water and wind (cross-bedded sands are a notable exception). It may represent a brief interruption of deposition, a short period of erosion, or a change in particle size or composition. For example, a flow of water depositing a thin layer of sand gradually slackens, leaving finer and finer sediments on its bed. Then a sudden renewal of flow once more deposits coarse materials, creating an abrupt transition from clay to sand.

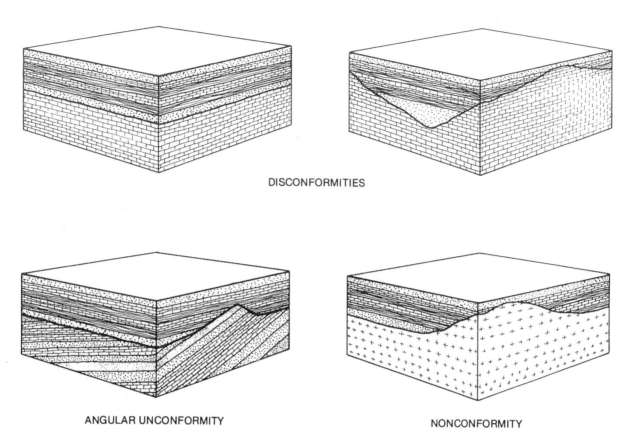

DISCONFORMITIES

ANGULAR UNCONFORMITY

NONCONFORMITY

Figure 2.46. Unconformities

Unconformities

A break in vertical continuity in the geologic record is known as an *unconformity* (fig. 2.46). An unconformity represents a time during which deposition did not take place or during which sediments or rocks formerly present were removed by erosion. It is a "fossil landscape"—a former surface of the earth, often complete with hills and valleys, which has been buried beneath later sedimentary deposits.

Disconformity. An unconformity along which the layers above and below are parallel is called a *disconformity*. The disconformity may be parallel to the layers, representing a period during which deposition simply ceased; or it may be irregular, indicating an episode of erosion. In the former case, the disconformity may be difficult or impossible to distinguish from an ordinary bedding plane. Unless revealed by correlation with nearby strata, a gap of thousands or millions of years of deposition may go undetected.

Angular unconformity. In an *angular unconformity*, the upper layers are not parallel with the lower ones. Older layers have been tilted or folded, eroded, then submerged for further deposition. An angular unconformity separates Precambrian sedimentary rocks from Paleozoic strata in parts of the Grand Canyon (fig. 2.47).

Nonconformity. A *nonconformity* is an unconformity in which sediments were deposited on an eroded surface of igneous or metamorphic rock. In this situation, the fundamental difference in character of the rocks above and below is usually obvious. However, if the formation below is of banded metamorphic rock, the nonconformity may

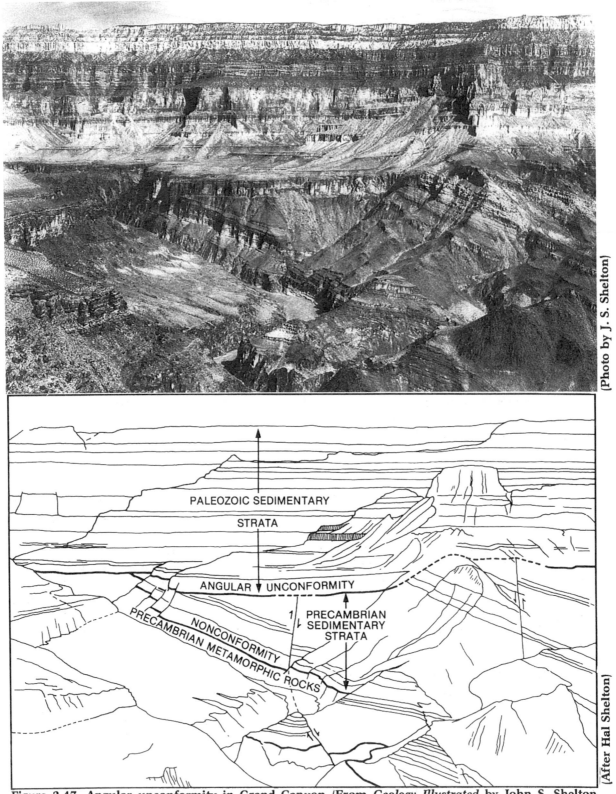

PALEOZOIC SEDIMENTARY

STRATA

ANGULAR UNCONFORMITY

PRECAMBRIAN
SEDIMENTARY
STRATA

NONCONFORMITY

PRECAMBRIAN METAMORPHIC ROCKS

Figure 2.47. Angular unconformity in Grand Canyon (From *Geology Illustrated* by John S. Shelton.
Copyright ©1966 by W. H. Freeman and Company. All rights reserved).

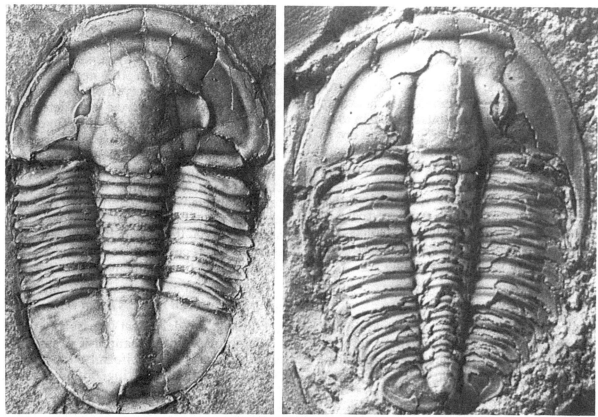

Figure 2.48. Fossil trilobites (Courtesy of A. R. Palmer, United States Geological Survey)

superficially resemble an angular unconformity.

Faunal Succession

Geologists use fossils to determine the relative ages of sedimentary rocks. Because both organisms and biotic communities evolve, fossils in a given area occur in a definite sequence. The absolute ages of fossils cannot be determined without using techniques such as radiometric dating, but the relative ages of most fossils are well known. For instance, trilobites (fig. 2.48) were most widespread and diverse early in the Ordovician period, about 500 million years ago. As the Paleozoic era progressed, however, their range and number of species declined, and they became extinct by the end of the Permian period, about 230 million years ago. Diatoms, on the other hand, appeared in the Cretaceous period, and their

glassy silica shells are found in rocks no older than about 135 million years (fig. 2.49). The governing principle is

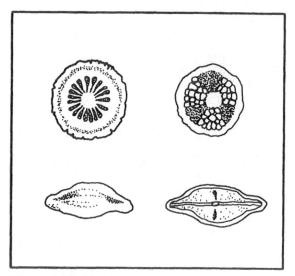

Figure 2.49. Diatoms

known as the law of *faunal succession:*

Fossil faunas and floras in a stratigraphic sequence succeed one another in a definite, recognizable order.

In other words, an extinct species cannot leave traces in rocks formed after its extinction. Any rock with trilobites is certain to be rock formed in the Paleozoic era and not in the Mesozoic or Cenozoic eras. In stratigraphic sequences that have been overturned by major tectonic movements, fossil assemblages are useful in showing that older rocks now overlie younger rocks, a condition that seemingly contradicts the law of superposition and that may not be obvious from casual observation.

Relative Age

As you can see, the story told by an assemblage of sedimentary rocks is often complicated by subsequent events. These events and the order in which they occur

are as much a part of the story as the origins of the rocks. It was not always apparent that a succession of events had affected a given area or that the order of these events could be discerned through the application of logic. By applying the principle of uniformitarianism, however, we can determine how such events were related in time.

Two other principles formulated by Steno are the doctrines of *original horizontality* and *original lateral continuity.* Sediments are deposited in essentially horizontal, continuous sheets. When sedimentary rock formations are tilted, folded, or broken into separate parts, the events that caused these disruptions occurred after, not before, the deposition of the sediments. Therefore:

Rock formations are older than any folds, faults, or erosional surfaces found within them.

Folding. Under great pressures, rocks undergo plastic deformation. If stress is applied gradually and uniformly, rock layers will bend without breaking, often becoming

Figure 2.50. **Folded strata in Canadian Rockies (Courtesy of Geological Survey of Canada, Ottawa, #180345)**

deformed into improbable shapes. In mountain ranges it is easy to find great folds in both sedimentary and other rock formations that have been caused by enormous horizontal and vertical forces (fig. 2.50). Rock folds that are convex upward (*against* the pull of gravity) are *anticlines;* folds that are convex downward (*with* gravity) are *synclines.* Synclines and anticlines range from a few feet to many miles across.

The orientation of a nonhorizontal rock layer is described in terms of its dip and strike. If a cannonball were placed on such a layer, the direction and downward angle that it rolled would be the *dip.* The line along which the plane of the formation intersects the horizontal plane is the *strike;* it is always perpendicular to the dip. A layer slanting down to the northeast might be described as having a dip of 20°NE and a strike of N45°W (fig. 2.51). On a folded formation, dip and strike vary from place to place. Dip is 0° at the crest of the anticline, increasing to either side of the crest. Notice that the strike does not change with distance from the crest.

A fold that has become slanted to one

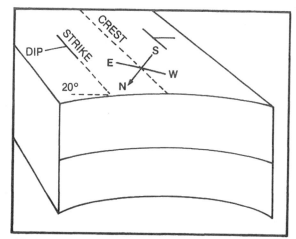

Figure 2.51. Dip and strike

side, so that the layers on one side occur in reverse order (younger layers beneath older) is an *overturned* fold (fig. 2.52). If the axial plane of an overturned fold is horizontal or nearly so, the fold is a *recumbent* fold. A fold whose long axis is not horizontal is termed a *plunging* fold.

Although a fold cannot be older than the rock formation it is found in, there are folded

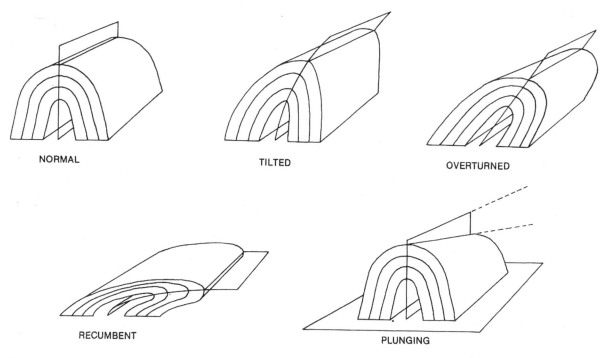

NORMAL TILTED OVERTURNED

RECUMBENT PLUNGING

Figure 2.52. Types of anticlines

INFILLING SYNCLINE

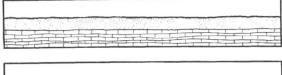

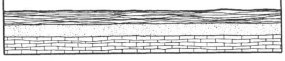

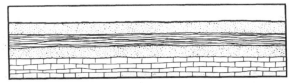

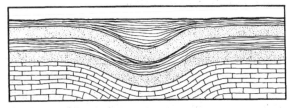

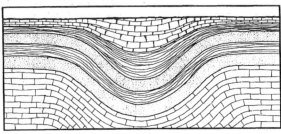

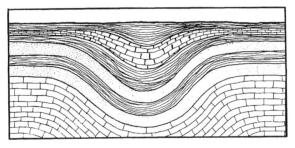

DRAPED ANTICLINE

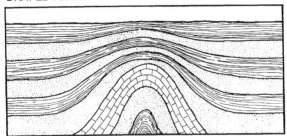

Figure 2.53. Infilling syncline and draped anticline

structures in which the deformation of the rock layers appears to have occurred at the same time the layers themselves were being formed. One example is an actively growing syncline that is continuously being filled with sediments (fig. 2.53). Such a feature is recognized by the tapering of its layers. The sedimentary deposits are thickest in the middle, where subsidence is most rapid. A *draped anticline*, conversely, is thinnest in the middle and thickest at the sides where subsidence is faster.

Faulting. A body of rock stressed beyond its breaking strength may become faulted. A *fault* is a break along which relative movement has occurred—that is, where the rocks on one side have become displaced relative to those on the other (fig. 2.54). This displacement can range from a fraction of an inch to hundreds of miles.

Displacement occurs along a *fault plane*, which is described in terms of strike and dip. Upward or downward displacement is called *dip slip;* horizontal displacement, as along the San Andreas fault in California, is *strike slip.* A combination of dip and strike slip produces *oblique slip.*

If the fault plane is not vertical—that is, if dip is less than 90°—the rock surface forming the underside of the fault is the *footwall*, and that on the upper side, the *hanging wall.* A dip-slip fault along which the hanging wall has moved down relative to the footwall is a *normal fault;* if the relative movement of the hanging wall has been upward, it is a *reverse fault* or *thrust fault.* A low-dip-angle (nearly horizontal) thrust fault along which a large displacement has occurred is termed an *overthrust fault.* Some overthrusts, including those in the Rocky Mountain Overthrust Belt, represent slippages of many miles.

A fault interrupts the continuity of rock strata; therefore, it cannot occur before deposition of the layers. However, deposition can continue after the fault begins to slip. As a result, the strata on the downthrown side are thicker than those on the other side. This type of fault, known as a *growth fault,* has a slip face much like that of a slump, or landslide (fig. 2.12).

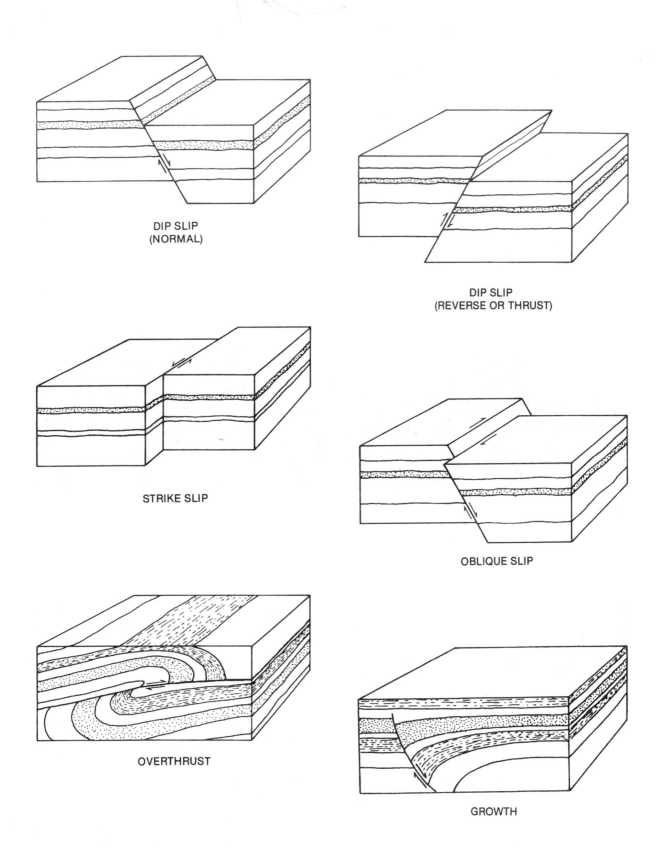

DIP SLIP
(NORMAL)

DIP SLIP
(REVERSE OR THRUST)

STRIKE SLIP

OBLIQUE SLIP

OVERTHRUST

GROWTH

Figure 2.54. Types of faults

Erosion. In Arizona's Monument Valley, only isolated "islands" remain of a once continuous layer of sandstone (fig. 2.55A). The rest has been removed by weathering and erosion. If this landscape, consisting mostly of horizontal strata, were buried beneath fresh sediments, a vertical cross section would show isolated bodies of sandstone beneath a disconformity (fig. 2.55B).

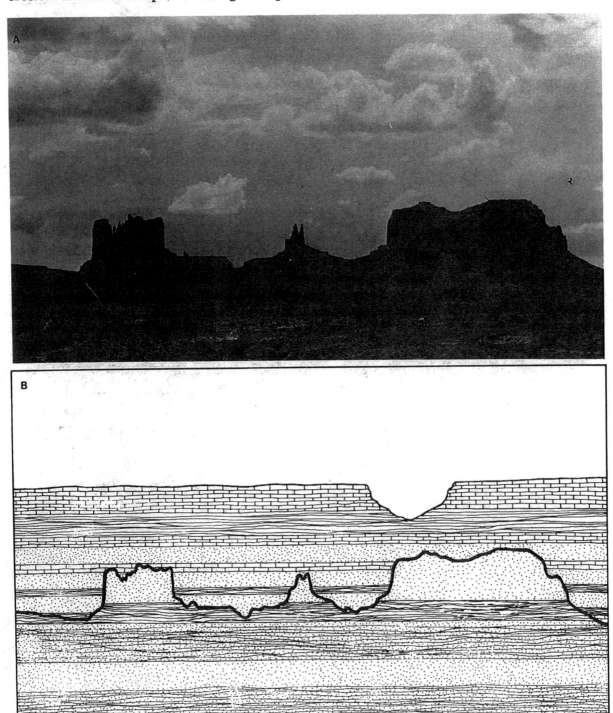

Figure 2.55. Monument Valley as outcrop, *A,* and disconformity, *B* (Photo by J. Morris)

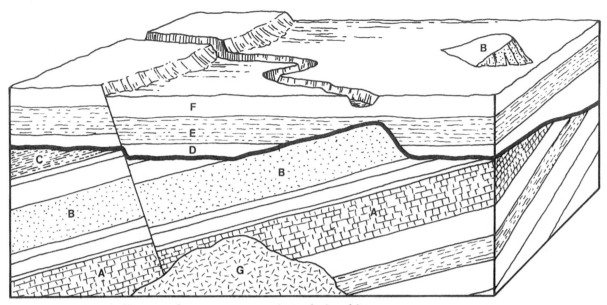

Figure 2.56. Block diagram showing cross-cutting relationships

Cross-cutting. Another concept that helps establish the relative dates of geologic events is the principle of cross-cutting relationships. Any feature that cuts across another feature must be younger than the feature it cuts.

In figure 2.56, the oldest features are the strata in the series topped by layer *A*. Series *B* is next oldest, followed by series *C*. The unconformity, which cuts across all three series, represents a time during which these layers were uplifted, tilted, and partly eroded. Next, the landscape represented by the unconformity was submerged—except for an island of outcrop *B*—and covered by the sediments of layers *D, E,* and *F.* Once again the area was uplifted, allowing the stream to begin downcutting into layer *F.* The fault, which cuts across all sedimentary layers, occurred during or after the regional uplift,

causing the stream to carve a deep canyon across the upthrown block. Finally, igneous rock *G* intruded into series *A.* This pluton was not cut by the fault and must therefore have occurred after the fault.

———

Of the three basic types of rock, sedimentary rock is unique in that it always forms as the result of processes acting on or near the earth's surface. Thus, the physical and chemical characteristics of sedimentary rock strongly reflect not only its source material but also the climate, geography, and biology in the area where it was formed. All of these factors—along with subsequent physical and chemical alterations—play a significant role in the formation and accumulation of fossil fuels.

3

OIL AND GAS ACCUMULATION

For most of their history, oil and natural gas were thought of as minerals, substances formed out of nonliving rock, just as gold, sulfur, and salt were part of the rock. There was little reason to assume otherwise. Although petroleum smelled like something that had died, and although natural gas burned like swamp gas, most of the gas and oil escaping from the ground seemed to come from solid rock deep beneath the surface, where, as everyone knew, nothing lived.

Beginning two centuries ago, however, the geologic insights of Hutton, Lyell, and other scientists showed that the rocks in which oil was found were once loose sediment piling up in shallow coastal waters where fish and algae and plankton and corals lived. Now it seemed possible that oil and gas had something to do with the decay of dead organisms, just as coal, with its leaf and stem imprints, seemed to be the fossilized remains of swamp plants.

Later advances in microscopy revealed that oil-producing and oil-bearing rocks often contain fossilized creatures too small to be seen with the unaided eye. Chemists discovered that the carbon-hydrogen ratios in petroleum are much like those in marine organisms and that certain complex molecules are found in petroleum that are otherwise known to occur only in living cells. But it was the fact that most source rocks could

be shown to have originated in an environment rich with life that clinched the organic theory of the origin of petroleum.

Unanswered questions about the occurrence of petroleum remain, and men of science still debate the evidence of its organic origin. Because of the weight of that evidence, however, few scientists doubt that most petroleum originates in the life and death of living things.

ORIGIN

Chemical Factors

Petroleum is both simple and complex. It is composed almost entirely of carbon and hydrogen; but the number of ways that carbon and hydrogen can combine is astronomical, and most petroleum contains hundreds of different kinds of hydrocarbons. It occurs in forms as diverse as thick black *asphalt* or *pitch*, oily black *heavy crude*, clear yellow *light crude*, and *petroleum gas*. These variations are due mainly to differences in molecular weight—that is, the sizes of the molecules—and the types of impurities. Despite the differences in molecular weights, however, the proportions of carbon and hydrogen do not vary appreciably among the different varieties of petroleum; carbon comprises 82 to 87 percent and hydrogen, 12 to 15 percent.

Petroleum is almost insoluble in pure water and only slightly soluble in salt water or water containing other organic substances. It is lighter than, and therefore floats on, water; but it is often found in an oil-water emulsion—that is, dispersed in small droplets suspended in water.

A hydrocarbon molecule is a chain of one or more carbon atoms with hydrogen atoms chemically bound to them. Some petroleum contains hydrocarbon molecules with up to sixty or seventy carbon atoms. At room temperature and pressure, molecules with up to four carbon atoms occur as gases;

molecules having five to fifteen carbon atoms are liquids; and heavier molecules occur as solids.

Methane, the simplest hydrocarbon, has the chemical formula CH_4 (fig. 3.1). Four is the maximum number of hydrogen atoms that can attach to a single carbon atom; thus methane is classified as a *saturated* hydrocarbon—a *paraffin*, or *alkane*. Other paraffins include *ethane* (C_2H_6), a chain of two carbon atoms with six hydrogen atoms; *butane* (C_4H_{10}); and *octane* (C_8H_{18}).

Unsaturated hydrocarbons also occur naturally in petroleum. The most common

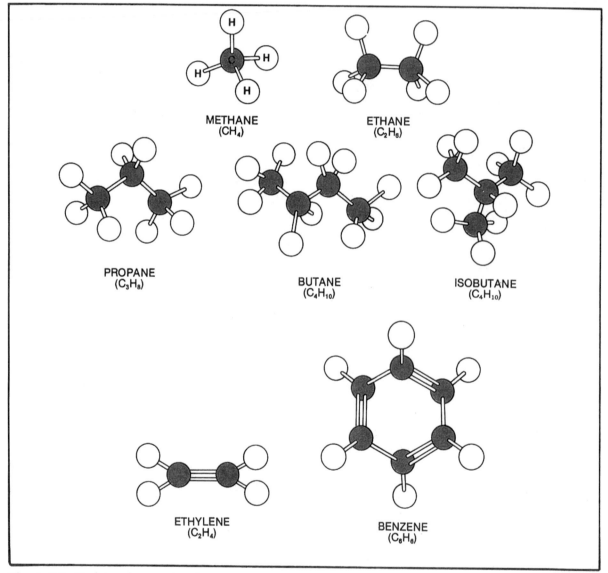

Figure 3.1. Hydrocarbon molecules

of these are the *aromatics*, compounds based on the distinctive *benzene ring* (C_6H_6). Other ring-shaped compounds include the *naphthenes,* or *cycloparaffins,* which vary in number of carbon atoms and bonding pattern. Naphthenic crudes produce less fuel and more lubricating oils than paraffinic crudes. The residue left when these crudes are refined is high in semisolid or solid asphaltic wastes; therefore, naphthenic oils are sometimes called *asphaltic* crudes.

Biological Factors

Petroleum contains solar energy stored as chemical energy. Many steps are involved in the conversion from the simple radiant energy of the sun to the complex molecules of hydrocarbons. Coastal waters, rich with nutrients brought in by rivers and upwelling deep-sea currents, support an elaborate community of organisms ranging from microscopic, single-celled plants and animals to large predatory fish and mammals (fig. 3.2). Some of the smallest and simplest of these organisms perform the first capture and conversion of the sun's radiant energy.

The bulk of the living matter in such biotic communities is in the form of microscopic or near-microscopic simple

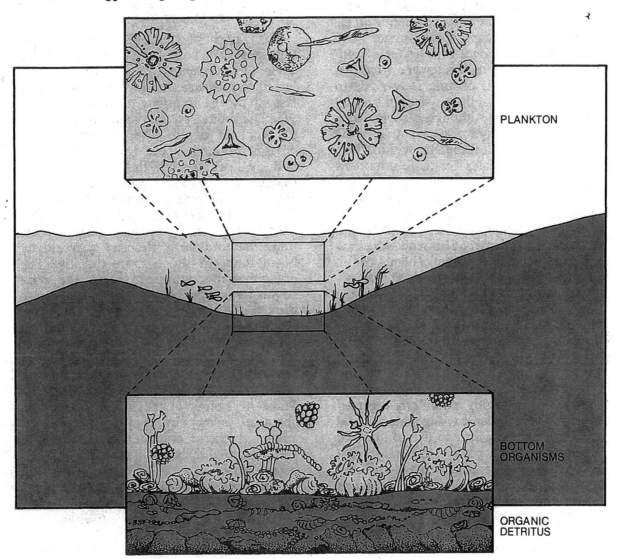

Figure 3.2. Shallow marine biotic community

organisms: protozoa (animals) and algae (plants). The algae are photosynthetic: they can synthesize their own food, simple sugars and starches, out of water and carbon dioxide, using the energy of sunlight. Other organisms consume the algae and convert the simple carbohydrates into more complex foods, such as proteins and fats; these, in turn, are eaten by still larger organisms.

Each level of the food chain contributes waste products, and every organism that is not eaten eventually dies. In recycling this organic material, an important role is played by a diversity of bacteria. The two principal types are those that live in *aerobic* (oxygenated) environments and derive their energy by oxidizing organic matter and those that live in *anaerobic* (reducing) environments by taking the oxygen from dissolved sulfates and organic fatty acids to produce sulfides (such as hydrogen sulfide) and hydrocarbons. Although aerobic decay liberates certain hydrocarbons that some small organisms accumulate within their bodies, the anaerobics are more important in the formation of oil.

If the process of aerobic decomposition continues indefinitely, all organic matter, including hydrocarbons, is converted into heat, water, and carbon dioxide—the raw materials that photosynthetic plants use to make their carbohydrate food. For an accumulation of petroleum to be formed, the supply of oxygen must be cut off. Most areas along the coast are well aerated by circulation, wind, and wave action. In some areas, however, aeration is hindered by physical barriers such as reefs or shoals; and in deeper waters far offshore, the water below a certain depth is similarly depleted of oxygen. Here organic waste materials and dead organisms can sink to the bottom and be preserved in an anaerobic environment instead of being decomposed by oxidizing bacteria. The accumulation and compaction of impermeable clay along with the organic matter help seal it off from dissolved oxygen. Thus isolated, it becomes the raw material that is transformed into petroleum by the heat and pressure of deeper burial.

Even in areas with appreciable circulation and oxygenation, organic debris can accumulate so fast that it is quickly buried beyond the reach of aerobic organisms. Locations where this is likely to occur include salt marshes, tidal lagoons, river deltas, and parts of the continental shelf. Epeiric seas such as those that covered much of North America during the Permian and other periods offered broad stretches of warm, shallow water where, unstirred by ocean currents and tides, abundant organic debris could accumulate in an anaerobic environment.

Physical Factors

The clay that settles out of suspension in quiet waters is buried and transformed into shale. Organic matter trapped within is subjected to pressure that increases at slightly less than the *geostatic pressure gradient*, which is about 1 pound per square inch (psi) per foot of depth. The temperature increases gradually, both from compression and by heating from the earth's interior. (Below a thin zone that is affected by climate, the temperature rises about 1.5°F for every 100 feet of depth.)

At 120° to 150°F, certain chemical reactions that ordinarily proceed very slowly begin to occur much more quickly. The organic matter trapped within the rock begins to change. Long-chain molecules are broken into shorter chains; other molecules are reformed, gaining or losing hydrogen; and some short-chain hydrocarbons are combined into longer chains and rings. The net result is that solid hydrocarbons are converted into liquid and gas hydrocarbons. Thus the energy of the sun, converted to chemical energy by plants, redistributed among all the creatures of the food chain, and preserved by burial, is transformed into petroleum.

The *petroleum window*—the set of conditions under which petroleum will form—includes temperatures between the extremes of 100°F and 350°F. The higher the temperature, the greater is the proportion of gas (fig. 3.3). Above 350°F almost all

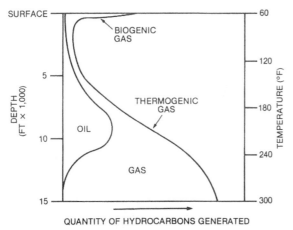

Figure 3.3. The petroleum window

of the hydrocarbon is changed into methane and graphite (pure carbon). Source beds (or reservoirs) deeper than about 20,000 feet usually produce only gas.

Source Rocks

Rock in which organic material has been converted into petroleum is called *source rock*. (Rock in which petroleum accumulates is called *reservoir rock*.) Generally, the best source rocks are shales rich in organic matter deposited in an anaerobic marine environment. Often these are dark shales, although the dark color can be caused by other substances. However, limestone, evaporites, and rocks formed from freshwater sedimentary deposition also become source beds.

Some petroleum geologists think that heat and pressure alone can convert organic

detritus to petroleum; others disagree on the relative importance of algae, plankton, foraminifera, and larger organisms in providing source organic material. It seems likely that petroleum is formed by a range of processes from a supermarket of raw materials under a variety of conditions; this fact would help account for the great chemical complexity of most petroleum and the variety of forms in which it occurs.

It takes time for petroleum to form and accumulate. Little petroleum has been found in Pleistocene formations or potential reservoir rocks associated with source beds less than a million years old.

MIGRATION

Like other formation fluids, oil and gas migrate. In some situations, they accumulate near where they originate, sometimes within a few inches or feet of the source bed. In other places, the migration covers many miles.

Because it is lighter than water and does not readily stay mixed with it, oil tends to separate from water and float on top. Usually it moves as a diffuse scattering of suspended droplets, but it may reach higher concentrations when its movement is impeded. Gas is usually present as well, either dissolved in the oil or as a separate, distinct accumulation.

The term *migration* is used in two senses (fig. 3.4). *Primary migration* is the movement of hydrocarbons out of the source rock—a

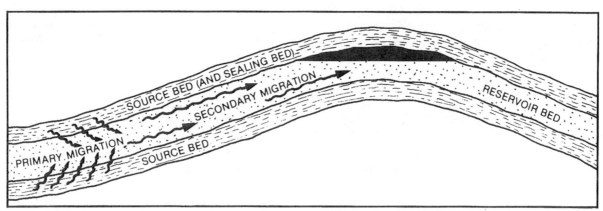

Figure 3.4. Primary and secondary hydrocarbon migration

journey of a fraction of an inch to several feet, rarely more. *Secondary migration* is the subsequent movement through porous, permeable reservoir rock by which oil and gas become concentrated in one locality.

Primary Migration

How does petroleum leave its source rock? Apparently both compaction and the flow of water are involved. Petroleum comes from source beds deposited mostly on the seafloor, so it usually begins and ends its travels in company with *interstitial water*—water found in the interstices, or pores, of the rock. (*Connate water*, a more exact term, is water that was "born with" the rock—present when the rock was formed.) Much more water than oil and gas is present underground.

As deposition continues at the surface, the growing weight of the overburden compresses the shale into less and less space. However, it is not the solid mineral grains that are compressed, but the pore spaces. Interstitial water is squeezed out, carrying droplets of oil in suspension and other hydrocarbons in solution. Although the solubility of oil in water is negligible compared to that of gas, both are more soluble under pressure.

Under compression, some rocks maintain their porosity and permeability better than others. Imagine two adjacent rock layers, a clean arenite sandstone and a silty shale, gradually being buried beneath thousands of feet of overburden. The sandstone will lose very little of its porosity because its relatively spherical grains were closely packed when deposited as sediments and its pores are not clogged with silt or clay (fig. 2.31*A*). Such a sandstone might have an initial porosity of 35%, which would be reduced to 30% at depth. Clay particles, on the other hand, are relatively irregular in shape and lithified in a loosely packed arrangement. Under pressure these particles become better aligned and more closely packed, like a pile of bricks rearranged to form a brick wall (fig. 2.31*B*). Under the

same compressive force as that on the sandstone, shale porosity might decrease from 60% to 35%. Fluids squeezed out of the shale will therefore collect in the adjacent sandstone, which retains more of its original porosity.

Secondary Migration

Hydrocarbons are moved through permeable rock by gravity. This force works in several ways: by compressing pore spaces containing fluid, by causing water containing hydrocarbons to flow, and by causing water to displace less dense petroleum fluids upward.

Saying that water flows through formations does not mean that it flows in underground rivers. Flow can mean movement of a few inches a year, which can add up to many miles in a geologically short time. What causes water to flow is a difference in *fluid potential* (which in some cases coincides with fluid pressure). Just as a difference in electrical potential causes electricity to flow from a high-voltage region to one of lower voltage, a difference in fluid potential causes water to move from a region of high fluid potential to one of lower potential.

On the surface, unconfined water moving from a high-potential to a low-potential zone simply flows downhill. In a municipal water system, however, water flows down from a high water tower, travels horizontally through a water main, and rises again into hilltop houses and the upper stories of high buildings (fig. 3.5). As long as its outlet is below the level of the water tower, water will flow uphill or down.

Water confined in a porous, permeable formation behaves much the same as water in a pipeline. In figure 3.6, water enters the formation at point *A* and exits at point *E*. A line drawn between these two points defines the *potentiometric surface*. (This is not the same as the *water table*, which is the upper surface of the underground water.) A well drilled into the aquifer at *B* will fill to point *F* on the potentiometric surface; an

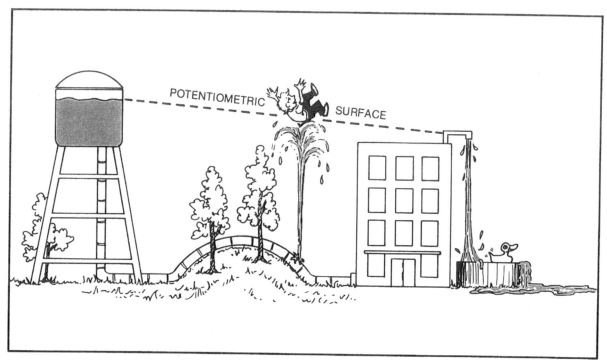

Figure 3.5. Municipal water system

artesian well drilled to *D* will seek level *H* above ground. Water flowing from *A* to *E* is flowing uphill from *B* to *C* and from *D* to *E*. It is not always flowing from high-pressure to low-pressure areas. Although it is doing so from *B* to *C*, the flow from *C* to *D* is toward greater hydrostatic pressure. The rate and direction of flow are the result of the difference in elevation between points *A* and *E*.

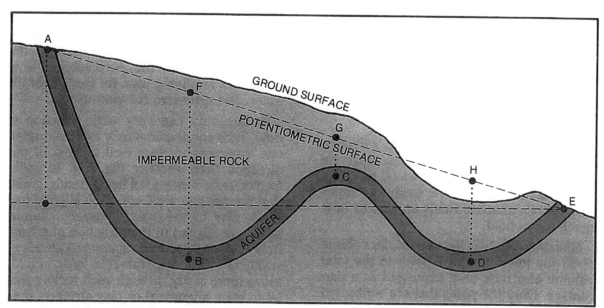

Figure 3.6. Potentiometric surface

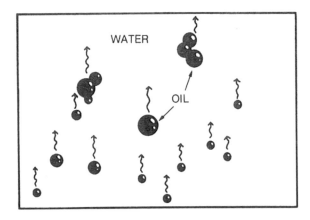

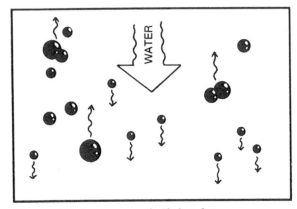

Figure 3.7. Buoyancy of oil droplets

Suspended oil droplets and dissolved gas are carried along in the flowing water. As oil saturation increases, however, small droplets coalesce into larger ones, and the accumulating oil begins to behave differently. Because of their buoyancy, large oil droplets tend to rise through water, often against a moderate flow, and accumulate at the top (fig. 3.7).

In secondary migration, the effective porosity and permeability of the reservoir rock are more important than total porosity. These factors control how easily the reservoir can accumulate fluids as well as how much it can hold.

Effective porosity is the ratio of the volume of all the interconnected pores to the total volume of a rock unit, expressed as a percentage. Only the pores that are connected with other pores (the effective porosity) are capable of accumulating petroleum. Effective porosity depends upon how the rock particles were deposited and cemented as well as upon later diagenetic changes. The structure of the rock may change in such a way that some of the pore spaces become isolated.

A rock's *permeability* is a measure of how easily fluids can pass through it. The basic unit of permeability is the *darcy;* $1/1000$ of a darcy is a *millidarcy* (md). Permeability is difficult to measure in the field because it varies greatly with pressure, fluid viscosity, oil saturation (the percentage of oil in the formation fluids), direction, and other factors. Depending upon sorting, compaction, cementation, and diagenesis, permeability can vary widely within a given type of rock. The permeability of sandstones commonly ranges between 0.01 and 10,000 md. Reservoir rock permeability of less than 1 md is considered poor, 1–10 md is fair, 10–100 md is good, and 100–1,000 md is very good. For comparison, a piece of writing chalk has a permeability of about 1 md.

Although often closely related, permeability and effective porosity are not the same (fig. 3.8). A rock with few isolated pore spaces may have a very high effective porosity and yet be nearly impermeable because of the narrowness of the connections between pores. Differences in *capillarity* (the ability of fluid to cling to the grains) may make the permeability of a given rock relatively high for gas, lower for water, and near zero for viscous oils.

Permeability can vary with direction of flow. Pore connections may be less numerous, narrower, or less well aligned in one direction than another (fig. 3.9). In rocks formed from well-sorted beach sands, grains that are not spherical are often aligned perpendicular to the beach. Stream channel sands are aligned in the direction of stream flow and often contain horizontal sheets or stringers of less permeable clay. Fluids move more easily through such rocks parallel to grain alignment or clay stringers than across them.

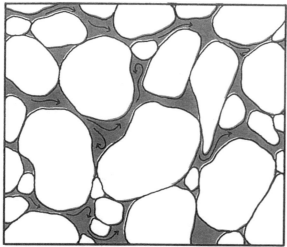

HIGH EFFECTIVE POROSITY, LOW PERMEABILITY

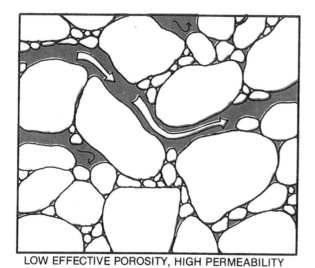

LOW EFFECTIVE POROSITY, HIGH PERMEABILITY

Figure 3.8. Permeability vs. effective porosity

ACCUMULATION

Like water in a puddle, oil collects in places it cannot readily flow out of. Its movement ceases upon reaching a structural high point or a zone of reduced permeability. As it accumulates, the mixture of hydrocarbons and water differentiates into distinct zones of water, oil, and gas.

Trapping

The permeability of the formation that seals off a petroleum reservoir is never absolutely zero, but just low enough to reduce the flow rate effectively to zero under reservoir conditions. Given enough pressure and fluidity (as opposed to viscosity), hydrocarbons may seep into a tight formation that under less extreme conditions would totally exclude them.

Effective permeability is the rock's permeability to a given fluid when another fluid is also present. Water has seven times the ability of oil to cling to the grains of porous rock, so it tends to fill small pores and keep oil out. In a petroleum reservoir, interstitial water is nearly always present. Clinging to the grains, it reduces the space available for oil and narrows the passages between pores, lowering the rock's effective permeability to oil.

PERMEABILITY 1,000 MD

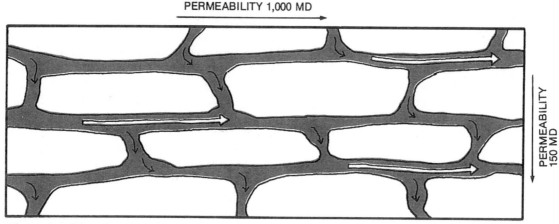

PERMEABILITY 150 MD

Figure 3.9. Directional variations in permeability

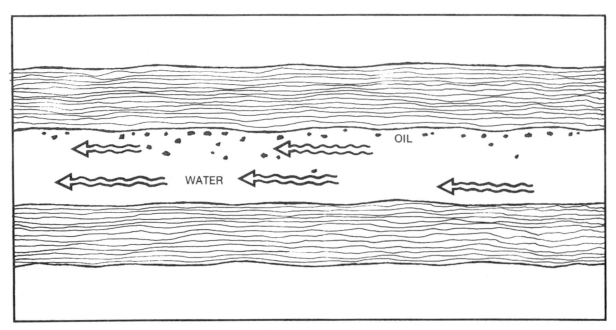

Figure 3.10. Hydrocarbon migration through horizontal formation

Relative permeability is the ratio of effective to absolute permeability. A rock that contains oil but no water has a relative permeability of 1.0 for oil. Relative permeability of 0.0 means that water has filled enough of the pore space to keep the oil from flowing. Higher fluid pressure may overcome the water's resistance to the passage of oil and may increase the effective permeability of the formation by opening tiny fractures across and along bedding planes.

A tight formation may keep fluids from leaving an underlying reservoir bed by preventing their vertical migration. However, fluids may still migrate horizontally beneath the seal. For an accumulation to form, petroleum fluids must encounter a *trap*, a geologic combination of impermeability and structure that stops any further migration. The basic mechanisms involved can be illustrated using as an example the anticline, the type of trap from which petroleum was first produced in commercial quantities and in which most of the world's presently known reserves are located.

In figure 3.10, interstitial water with a small concentration of oil droplets and dissolved hydrocarbons is coursing slowly through an undeformed horizontal sandstone beneath a layer of relatively impermeable shale. The oil droplets, although swept along by the movement of the water, tend to concentrate in the upper levels of the sandstone because of their buoyancy.

Suppose these horizontal layers are tilted by tectonic forces (fig. 3.11). If the tilt is upward in the direction of flow, the situation is little changed; oil tends to rise updip, the same direction in which the water flows. However, if the water flows downdip, the oil tends to rise against the flow of water.

Now suppose this sedimentary rock assemblage is bent into an anticline (fig. 3.12). Water flows through the permeable sandstone as before—updip into the anticline and downdip out of it. Oil rises with the water entering the anticline—but rises against the flow on the downdip side. If the water is not flowing too fast, the oil droplets brought in by the flowing water move preferentially toward the crest of the anticline. They concentrate and coalesce near the highest point. As more oil is brought in by the water, the pool grows.

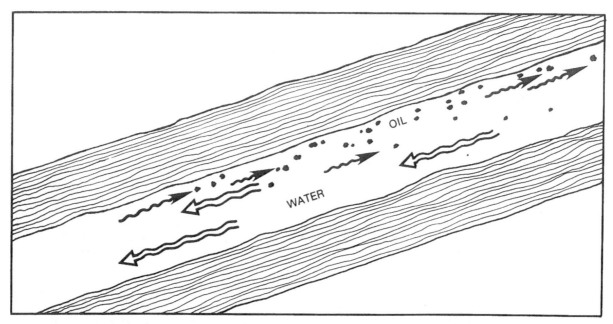

Figure 3.11. Hydrocarbon migration against flow in tilted formation

To accumulate oil, the anticline must be closed: it must dip toward both flanks and plunge in both directions along its axis.

Otherwise the oil will continue to migrate updip. An anticline is like an upside-down trough (or cup) out of which oil spills

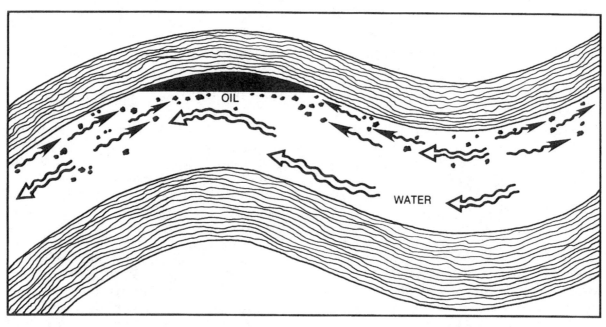

HYDROCARBON MIGRATION

WATER FLOW

Figure 3.12. Hydrocarbon migration into anticline

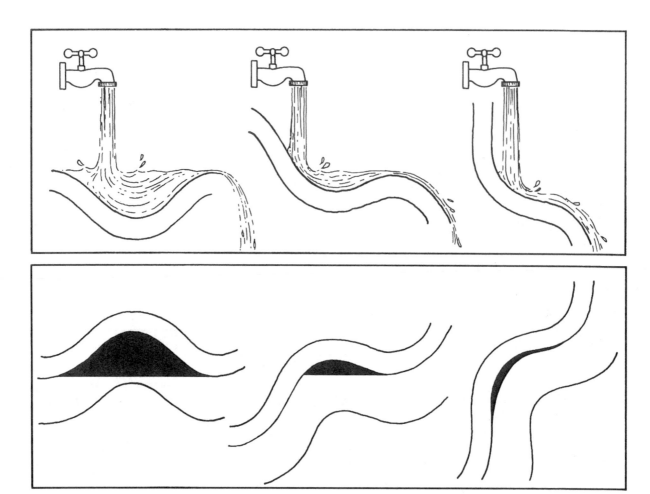

Figure 3.13. Analogy between pooling water and accumulating hydrocarbons

upward rather than downward. If the cup is tilted too much, the accumulated hydrocarbons will pour out of the cup and continue rising (fig. 3.13).

Differentiation

With few exceptions, petroleum reservoirs are water-wet—that is, the oil is not in contact with the rock grains because they are coated with a film of water. Most oil fields have 50% to 80% maximum oil saturation. Above 80% oil saturation, the oil can be produced with very little water mixed in; below 10%, the oil is not recoverable.

A hydrocarbon reservoir is divided into two or more zones. If only oil and water are present, the oil occupies the upper zone (fig. 3.14*A*). Although water still lines the

pores, this is the zone in which maximum oil saturation occurs. The oil zone is underlain by water along the *oil-water contact*, which is not a sharp line but a transition zone usually many feet thick. Oil saturation increases gradually from near 0% at the base of this zone to 50% to 80% at the top. The region of maximum oil saturation extends from the top of the transition to the top of the reservoir.

Natural gas is present in nearly all hydrocarbon reservoirs, dissolved in the oil (as *solution gas*) and to some extent in the water. In many situations, however, reservoir conditions and gas saturation allow undissolved gas (*associated free gas*) to accumulate above the oil zone as a *gas cap* (fig. 3.14*B*). The wetting fluid in a gas cap is usually water but occasionally oil. The transition zone between

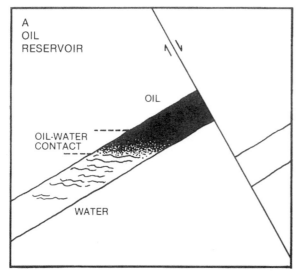

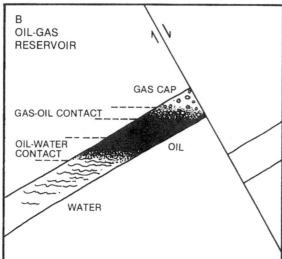

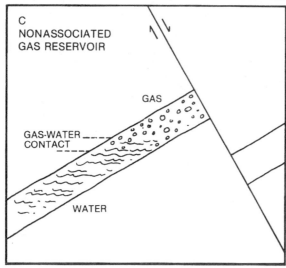

Figure 3.14. Basic hydrocarbon reservoir types

oil and gas (the *gas-oil contact*) is thinner than the oil-water contact zone because of the greater difference in density and surface tension between gas and oil.

Some reservoirs contain gas but not oil (fig. 3.14C). This gas is called *nonassociated gas*. The transition zone is a *gas-water contact*. The water almost always contains gas in solution, and water lines the pores in the gas zone.

Free methane, the lightest hydrocarbon, remains in a gaseous state even under great pressure. Ethane, propane, and butane, which are gases at surface pressure and temperature, are often found in a liquid state under reservoir conditions.

TYPES OF TRAPS

The basic requirements for a petroleum reservoir are a source of hydrocarbons, a porous and permeable rock formation through which hydrocarbons can migrate, and something to arrest the migration and cause an accumulation. Although most of the known reservoirs are anticlinal, many other geologic situations can cause hydrocarbons to accumulate.

Petroleum geologists have, for convenience, lumped hydrocarbon traps into two major groups: *structural,* in which the trap is primarily the result of deformation of the rock strata; and *stratigraphic,* in which the trap is a direct consequence of depositional variations that affect the reservoir formation itself. Some geologists also include a third type—*hydrodynamic* traps, in which the major trapping mechanism is the force of moving water. Most reservoirs have characteristics of more than one type.

Structural Traps

Anticlines. Anticlinal reservoirs are created by tectonic deformation of flat-lying and parallel rock strata. The basic anticlinal trap has already been described, and the syncline is not a noteworthy type of trap because only under rare circumstances does it cause petroleum to accumulate.

A short anticline plunging in both directions along its strike is classified as a *dome*. A dome is distinguished on structural maps by its nearly circular shape (fig. 3.15). Many domes are the result of *diapirism*, the penetration of overlying layers by a rising column of salt or other light, easily deformed mineral. Salt domes are common along the U.S. Gulf Coast and in the Middle East, northern Germany, and the Caspian Volga area of Russia.

Figure 3.15. Structural map of dome (From R. I. 140 by Martin Jackson and Steve Seni, Bureau of Economic Geology, The University of Texas at Austin)

Faults. When deformational forces exceed the breaking strength of rock, the result is a fault. Faults can affect the migration and accumulation of petroleum in various ways. They can act as conduits for water, oil, and gas, allowing or forcing these fluids into formations at other levels or even to the surface. Conversely, the movement of an active fault may grind rock into a fine-grained substance called *gouge,* which may be nearly impermeable to fluids. Shale strata cut by a fault may be smeared along the fault zone, creating a low-permeability barrier. A fault containing material of low permeability is called a *sealing fault.*

Most faults trap oil and gas by interrupting the lateral continuity of a permeable formation. Fault displacement places an impermeable formation opposite the reservoir rock, and updip hydrocarbon migration is stopped as suddenly and effectively as though a dam had been erected (fig. 3.16).

Faults and folds are often closely associated, and each can affect how the other causes petroleum fluids to accumulate. The

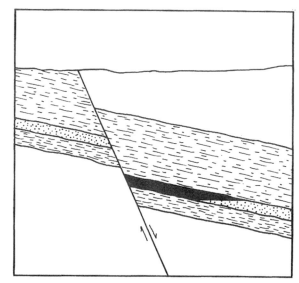

Figure 3.16. Fault trap

upper parts of anticlines are commonly faulted (fig. 3.17). Such faults, which often occur as the anticline is forming, may not change the total volume of the reservoir but may divide it into separate compartments.

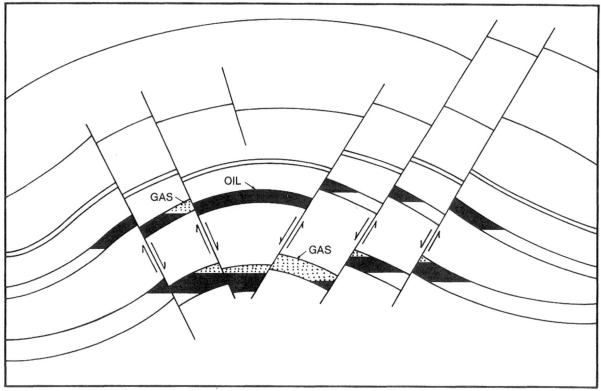

Figure 3.17. Faulted anticline traps

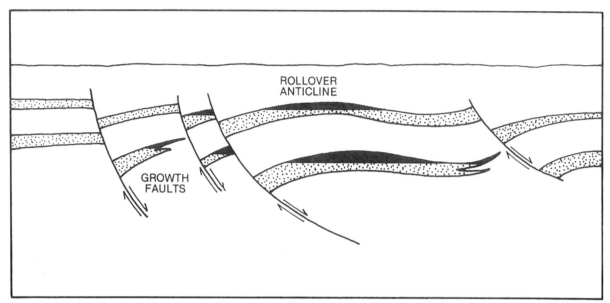

Figure 3.18. Rollover anticline and growth fault traps

Sealing faults may isolate the compartments from one another; nonsealing faults may allow reservoir fluids to redistribute themselves or even migrate out.

The dip of a growth fault approaches the horizontal at depth; deposition is faster on the downthrown side, which tends to "roll over" or curl downward (fig. 3.18). The result is a *rollover anticline,* where oil and gas can accumulate. Although not trapped by the fault itself, the petroleum may be trapped as an indirect result of it.

Great horizontal stresses, such as those caused by the collision of two continental masses, may be relieved through overthrust

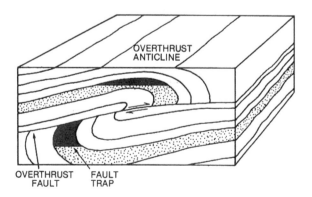

Figure 3.19. Overthrust fault traps

faulting. Blocks of the crust are shoved horizontally, sometimes many miles, atop their counterparts, creating a double thickness (or more, in complex cases) of sedimentary strata (fig. 3.19). The resulting flexures and faults can trap hydrocarbons in several ways. The crest of an overthrust anticline, sometimes called a *drag fold,* can become a reservoir, as can the sheared-off permeable rocks bent upward against the lower side of the fault. The Rocky Mountain Overthrust Belt contains many such reservoirs.

Stratigraphic Traps

Compared to anticlines and faults, traps that are the result of lateral discontinuity or changes in permeability are difficult to detect. For this reason, stratigraphic traps were not well represented or studied until long after many of the world's most productive structural oil fields had been discovered. The search for stratigraphic traps has intensified in recent decades, but they still account for only a minor part of the world's known petroleum reserves.

Many structural traps, especially those near the surface, can be located by careful study of the surface geology. Stratigraphic traps, however, are usually unrelated to surface features. Their elusiveness has stimulated the development of more and more sophisticated exploration techniques and devices. Many stratigraphic traps have been discovered accidentally while drilling structural traps.

Shoestring sand. A good example of why stratigraphic traps are hard to find is the *shoestring sand* (fig. 3.20). This type of reservoir is often an overlapping series of coarser stream sediments that were inundated and buried beneath thick deposits of clay. It appears as a sinuous string of sandstone winding erratically through impermeable shales. Multiple shoestring sands sometimes form complex branching networks. The overlying land surface, whether flat or cut by deep valleys, gives no hint of the underlying buried channel. Tracing the course of a shoestring is almost as difficult as finding it. The petroleum geologist uses such clues as

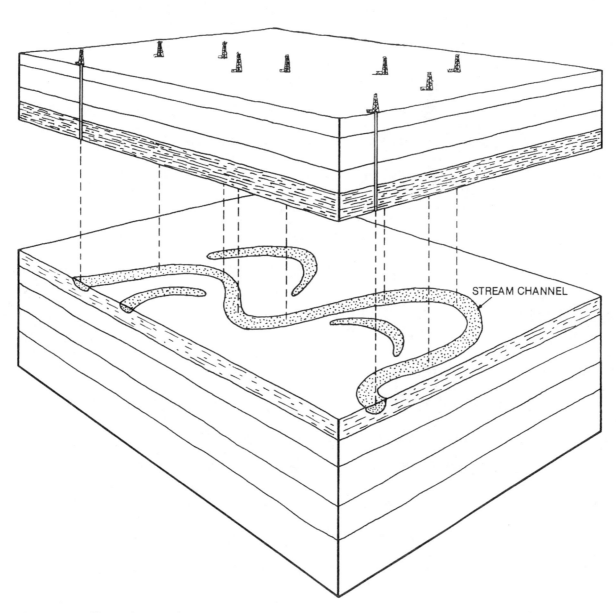

Figure 3.20. Shoestring sand

direction of greatest permeability and general slope of the buried land surface to find the next productive location. Lining up two or three productive wells helps narrow the search.

Beach sand. The linear zone of sediments marking a former coastline may form a series of sandstone reservoirs (fig. 3.21). Unlike river sandbars, grain orientation and direction of maximum permeability are *across* the trend. Information on facies changes is especially useful in locating this

type of reservoir. Knowing that limestone and shale usually form some distance offshore, the geologist can determine the direction in which the sediments become coarser and more permeable. The coarsest, best sorted, most permeable sand is found in the shoreline zone that was affected by wave action. It is the most likely place to find oil, both because of its porosity and permeability and because, if tectonically undeformed, it is structurally the highest part of the stratigraphic unit.

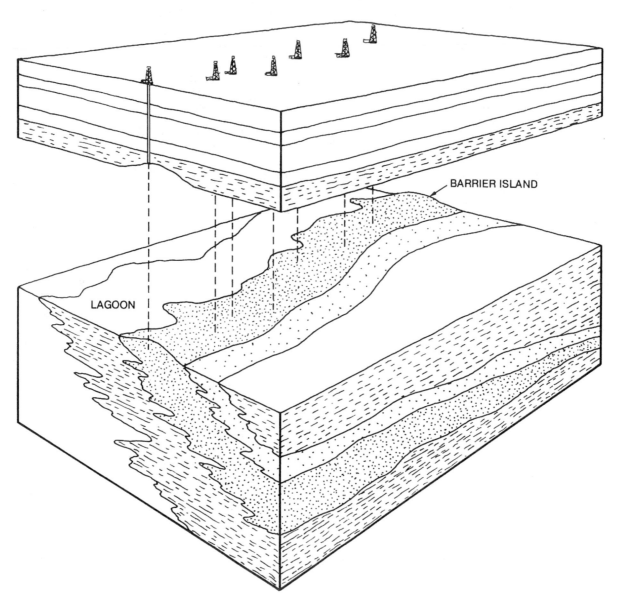

Figure 3.21. Bar sand

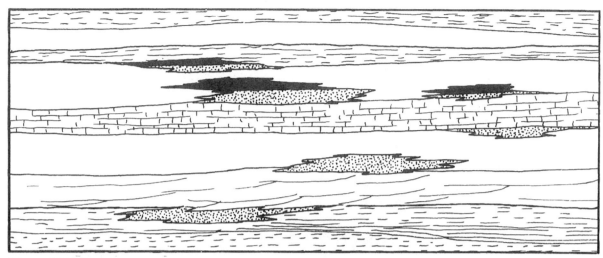

Figure 3.22. Lens traps

Lens. A *lens* is an isolated body of sandstone or other permeable rock enclosed within shale or other less permeable rock (fig. 3.22). Its edges taper out in all directions. Lenses can be formed by turbidity currents, underwater slides, isolated beach or stream sand deposits, alluvial fans, and other deposits. They are tapered in cross section, like shoestring and beach sandstones, but not extended in length.

Bioherm. Another type of stratigraphic trap is the reef (fig. 3.23). A wave-resistant accumulation of coral or shells serves as an anchor for calcareous debris that forms limestone. If deeply submerged faster than it can accrete, it may become buried beneath marine shales. Such an isolated reef is called a *bioherm*. It may be porous enough to hold large accumulations of hydrocarbons, especially if it has been dolomitized. Limestone is especially vulnerable to dissolution by groundwater, particularly if raised above the water table. Leaching by weak solutions of atmospheric carbon dioxide may form *vugs* (small voids) or *caverns* (large voids) capable of containing hydrocarbons. Overlying deposits may be laid down in such a way as to form a draped anticline, or *compaction anticline*—a structural feature that may also trap oil and gas. Atolls (rings

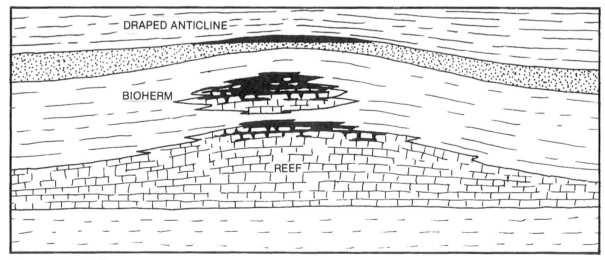

Figure 3.23. Bioherm

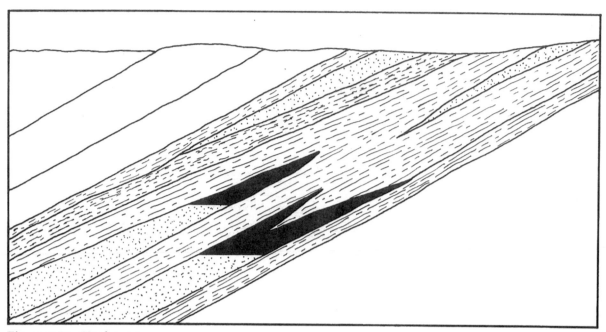

Figure 3.24. Pinchout traps

of coral islands), coral pinnacles, and other reef features are prolific oil producers. The Horseshoe Atoll, a buried shell reef system in West Texas, contains one of the world's largest and most prolific oil reservoirs (fig. 2.26).

Pinchout. A pinchout trap occurs where a porous and permeable sand body is isolated above, below, and at its updip edge by shale or other less permeable sediments (fig. 3.24). Oil or gas enters the sand body and migrates updip until it reaches the low-permeability zone where the reservoir "pinches out."

Permeability changes. Diagenetic changes can create traps within formerly permeable rocks or can create reservoir areas within formerly impermeable rocks. Minerals crystallizing out of circulating water between the grains of a porous sandstone may reduce local permeability enough to form a barrier to hydrocarbon migration. Alternatively, circulating water may increase permeability by leaching out cement or by enlarging fissures and vugs in limestone, thereby increasing the potential for hydrocarbon accumulation.

Fine-grained rocks such as the Austin Chalk in Texas may be unsuitable as reservoirs because of their relative impermeability. When cut by a fault, however, they may become locally fractured (brecciated) and accumulate small but productive pools in areas that are not structurally high (fig. 3.25).

Petroleum itself can seal permeable rock. If exposed to oxygen or altered by bacteria,

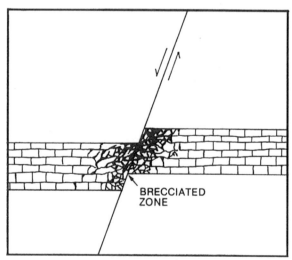

Figure 3.25. Brecciated fault trap

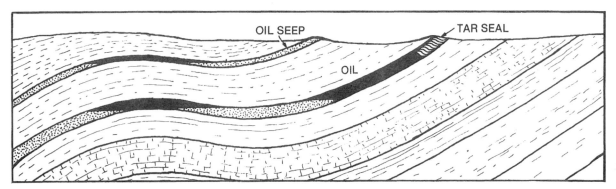

Figure 3.26. Asphalt seal trap

an oil seep often loses its more volatile components and becomes thick and tarlike. It may plug the rock pores so tightly that no further migration toward the surface can occur (fig. 3.26). Below the affected zone, however, recoverable petroleum may still accumulate.

Unconformities. A major subcategory of stratigraphic traps includes those associated with unconformities (fig. 3.27). Permeable outcrops overlain by impermeable layers can accumulate hydrocarbons from sources both above and below the unconformity. A buried anticlinal outcrop may trap petroleum in both flanks, downdip from the eroded crest. Locations that would have been oil seeps on the former landscape surface become traps beneath an unconformity. A porous, permeable formation lapping out on an unconformity can also become a trap (fig. 3.28).

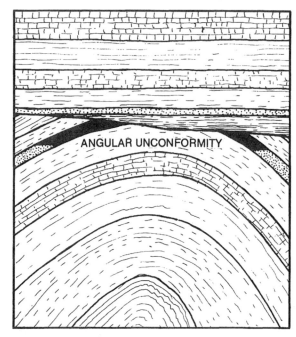

Figure 3.27. Unconformity traps

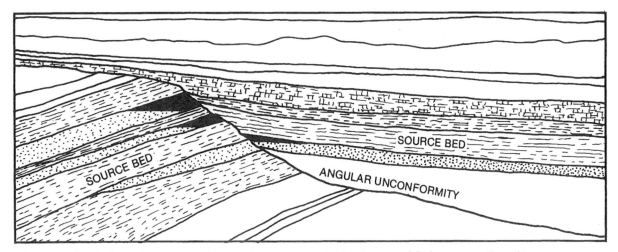

Figure 3.28. Hydrocarbon accumulation from source beds above and below unconformity

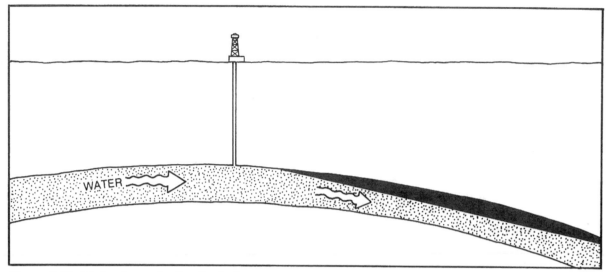

Figure 3.29. Hydrodynamically tilted oil-water contact

Hydrodynamic Traps

Movement of water through reservoir rock affects not only the amount but also the distribution of oil. An oil-water contact is usually tilted downward in the direction of flow. The pool can be displaced so far that a well drilled into the crest of an anticline might produce only water (fig. 3.29). The slope of the oil-water interface, and therefore the location of the pool, is related in a predictable way to the slope of the potentiometric surface (which is always in the direction of flow) and to the difference in density between the oil and the water. The denser the oil, the more it is displaced. Gas,

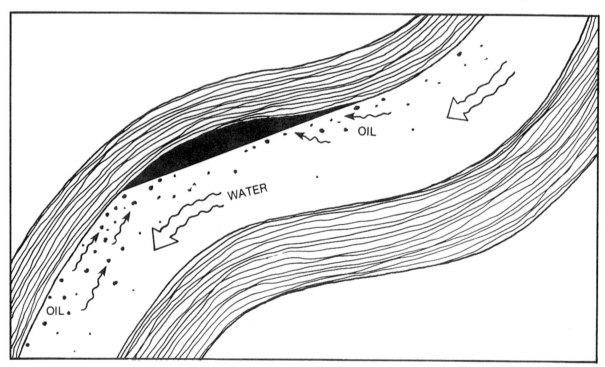

Figure 3.30. Hydrodynamic trapping of oil in tilted anticline

the lightest petroleum fraction, tends to stay near the crest of the anticline.

Oil may accumulate hydrodynamically in a structural feature that might otherwise not trap it. The flow of water through the reservoir bed in figure 3.30 causes oil to accumulate in the tilted anticline. Were water flow to cease, the buoyancy of the oil would cause it to migrate updip.

Water flows through a confined permeable layer at a definite volumetric rate. If the cross-sectional area varies, the volumetric flow rate remains constant, but the linear flow rate changes. In other words, to maintain the same number of gallons per hour, the water must flow faster through a narrow section than through a broad section.

In figure 3.31, water flows downdip from a narrow zone into a broader section. Buoyant oil droplets can migrate upstream against the slow flow in the broad section but not against the rapid flow in the narrows. Oil therefore backs up in a pool at this bottleneck. A zone of reduced permeability can have the same effect because the cross-sectional area of the interconnected pore space is reduced.

Combination Traps

Many petroleum traps have both stratigraphic and structural features. Some, in which both types of characteristics are essential in trapping petroleum, are difficult to classify as either primarily structural or primarily stratigraphic. For instance, originally horizontal formations that now pinch out updip can trap hydrocarbons that might not otherwise have accumulated. Secondary porosity in a shattered (brecciated) fault zone or anticlinal crest is a stratigraphic trapping mechanism caused by structural deformation. Most hydrodynamic trapping depends partly upon formation structural features and often upon stratigraphic variations within the reservoir formation.

Many types of traps can be found near salt domes. Most would be considered structural, although some could be classified as combination traps. Beneath the U.S. Gulf Coast are thick beds of salt that were deposited, during the opening of the modern Atlantic Ocean, in sedimentary basins with restricted circulation and high evaporation rates. Under pressure, this rock salt, light and easily deformed, is displaced by the weight of accumulating sediments, forming huge mushroomlike columns that rise

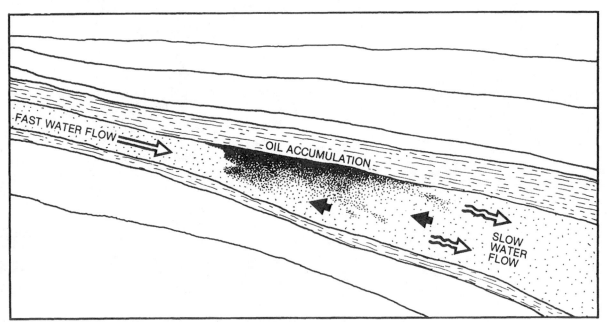

Figure 3.31. Hydrodynamic trapping of oil in wedge-shaped formation

toward the surface. In each of these *dia-pirs,* the overlying rocks are pushed aside or bulged upward into a dome (fig. 3.32). The penetrated layers are dragged upward by the rising salt core and depressed downward away from it as the overlying layers subside to replace the depleted salt bed. Leaching by groundwater prevents the salt from breaking through the surface, but leaves atop the column a residue of less

soluble compounds, forming a dense, nearly impermeable *caprock.* Overlying domed sediments break in a complex series of intersecting faults. The base of the salt core may narrow, creating a mushroom-shaped overhanging column.

Many types of petroleum traps are thus formed: a multilayered dome on top, cut by faults; upturned drag folds that terminate against impermeable salt; upturned pinch-

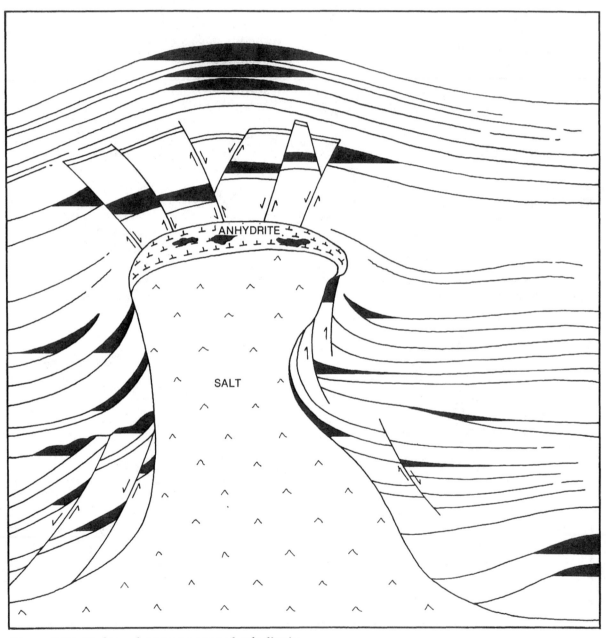

Figure 3.32. Hydrocarbon traps around salt diapir

outs where compression and other diagenetic changes have reduced permeability; and faults along the flanks. Oil may also collect beneath the impermeable caprock or beneath the overhanging salt. The multiple possibilities for traps and the high likelihood of finding petroleum have made salt domes popular places to drill.

TIMING AND PRESERVATION OF TRAPS

The accumulation of oil underground is a dynamic phenomenon. It is a function not only of the location and configuration of source and reservoir beds but also of timing – the formation of the trap and the arrival of the resource. A potential trap may form but remain barren because all the petroleum has migrated out of the area; or it may form and be destroyed before hydrocarbons are generated. An accumulation of hydrocarbons in a perfectly suitable reservoir may eventually be lost either because of deformation and destruction of its trap or by excessive heat and pressure.

As an example of how the amount and character of hydrocarbon accumulation may change over both time and space, consider the stairstep sequence of anticlines in figure 3.33. Water, with entrained oil and

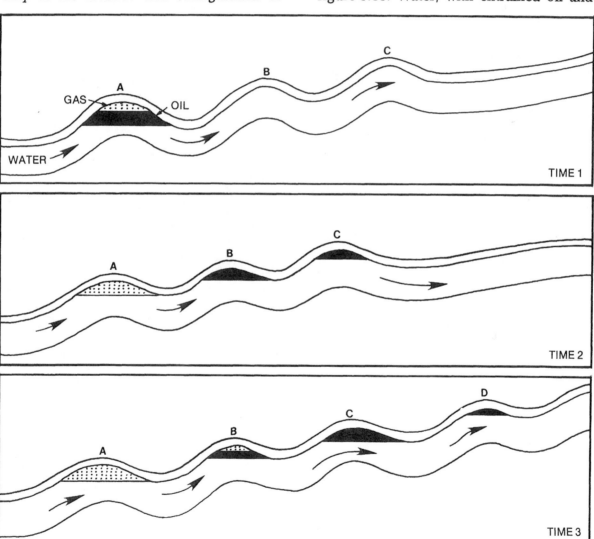

Figure 3.33. **Migration of hydrocarbons through anticline series**

gas, flows from left to right. At time *1,* oil and gas are trapped in anticline *A* while "cleaned" water continues through *B* and *C.* As more oil and gas accumulate, the trap fills and some of the oil spills out into trap *B.* The gas, however, continues to accumulate in *A* until at time *2* it has displaced all the oil. The process continues, as at time *3.* Thus, the range of fluids produced by a well in any of these structures depends upon the stage in its history.

An old trap may contain new oil—that is, oil that originated long after the trap was formed. Conversely, a new trap may contain oil that is older than itself. Suppose that anticline *D* in figure 3.33 forms between times *2* and *3.* The "old" oil spilling from anticlines *A, B,* and *C* can then accumulate in the new trap, *D.*

Usually there is some correspondence between the age of the trap and the age of the oil, because source and reservoir rocks tend to be fairly close together and subject to similar stresses. The stress that deforms a reservoir bed to produce a petroleum trap may also bring about the generation and primary migration of hydrocarbons from nearby source rocks.

Whether or not an accumulation of hydrocarbons is preserved depends upon many factors. Fissuring or a sealing formation under bending stress; changes in porosity and permeability of limestone due to leaching or dolomitization; excessive heat and pressure imposed on a reservoir beneath accumulating sediments—these are a few of the many circumstances in which petroleum that was generated through most of the history of the planet has failed to survive to the present. The oil and gas produced today are only remnants of a much greater resource, most of which has disappeared.

Spared the effects of excessive heat and pressure, bacterial decomposition, diagenetic destruction, and uplift and erosion, a pool of petroleum may endure for many millions of years. Most of the world's known oil reserves are found in Mesozoic formations (65 to 225 million years old); smaller amounts are found in Paleozoic rocks (up to 500 million years old) or Cenozoic rocks (less than 65 million years old).

————

The petroleum geologist's job is finding oil and gas that can be produced for commercial profit. Most of what he needs to know in the performance of his duties is hidden beneath the surface. But he has access to powerful tools and techniques for revealing the secrets of the earth's crust, and special ways of analyzing this information that enable him to find the best places to drill exploratory wells and to help his company make risk decisions. To the geologist's trained eye and mind, the same rocks that hide the resource also provide subtle clues about its location.

4

EXPLORATION

The search for minerals has not only powered the industrial revolution but also helped us learn more about the planet we live on. To use a recent example, geophysical surveys conducted in North America by companies exploring for iron, copper, and petroleum have revealed the presence of a great subsurface continental rift extending from the Great Lakes to the Rocky Mountains. This discovery seems certain not only to add to our known reserves of oil but also to provide insight into how the global forces of plate tectonics created and changed this continent—helping, in turn, to direct the continuing search for resources.

The petroleum geologist's main job is to identify locations most likely to produce oil and gas. In accomplishing this objective, he will at various times find himself examining data that have been gathered by others while exploring or drilling; conducting, directing, or contracting for geophysical or geological surveys to obtain new information; constructing, using, and adding to maps, sections, and other graphic data to document a prospect; and evaluating the commercial potential of a prospect. His contributions and conclusions not only make possible the cost-efficient location and production of petroleum reserves but also add to the store of information about our planet.

LOCATING THE PROSPECT

In exploring for locations where oil and gas might be found, the petroleum geologist starts from the known and works toward the unknown. Most oil-producing areas (*provinces*) have been thoroughly mapped; the geologist working in these areas can draw on an abundance of both raw and compiled data to locate a *prospect*—a new area where he thinks oil might be found.

Using Data Libraries

Much of the information that a company geologist works with is proprietary, that is, owned by his company and not shared with competitors. Other data are available to everyone, either free or for a price.

Libraries of geologic and oil field data are invaluable to petroleum geologists. By gaining access to work already performed, the geologist can get a head start in gathering data for new leases. An oil company may maintain a data library for all areas that it has explored or developed, but such resources rarely hold all that is known about a particular area. The geologist may obtain much additional information from university well log libraries, public agencies, private companies, and databases.

Public agencies. Public agencies receive many kinds of data from oil companies that drill leases in the United States. The Railroad Commission of Texas and the Oklahoma Corporation Commission are examples of agencies that regulate oil and gas industry activities at the state level. State regulations specify the types of data to be collected and the steps for reporting them. The files are open to the public and accessible to anyone who knows the name of a well, the county where it was drilled, who drilled it, and when it was drilled.

Private companies. Many well log libraries are private companies or geological societies that maintain large collections of well logs, scout tickets, well test reports, completion cards, production data, and maps. Members pay dues for access to the information, which the libraries obtain both from public agencies and from members who contribute their own data. Some log libraries offer map services. Small oil companies are heavy users of well log libraries, whose price of membership is often less than the cost of maintaining company files.

Databases. Both public and private organizations have established computer databases that offer access to a variety of mostly regional information, including reservoir geology, engineering, and production data, classified by field and reservoir. Major oil companies subscribe to petroleum databases. Nonsubscribers may request service on a per-search basis. As the use of computers increases, the petroleum industry will undoubtedly make greater use of public and private database services.

Exploring within Known Oil Provinces

A petroleum geologist usually works from *base maps*—charts showing existing wells, property lines, roads, buildings, and other surface geographic features. Sometimes, as in a partly developed field, the main factor in his recommendation for a new drilling site is simply a choice among existing leases. When considering sites away from the main drilling activity, he may recommend further exploration by seismic or other methods to help determine where leases should be sought.

Structure maps of critical horizons—subsurface oil-producing zones—are an important part of the geologist's information. A formation that extends over a broad area, such as the Ellenburger in West Texas, may be a known oil producer in some locations but a nonproducer in others. The geologist looks for localities where formation characteristics are favorable to the accumulation of petroleum. He studies well logs, cores, and other locally obtained data for leads—clues that might help narrow the search for a prospect.

Exploring Unknown Areas

In an unexplored area, the geologist must begin his search by gathering basic information. Surface topography and near-surface structure, as well as geographic features such as drainage and development, can be studied through aerial and satellite photography. Regional and deep subsurface structure can be broadly outlined by using relatively inexpensive magnetic and gravimetric surveys; interesting localities can be brought into sharper focus by seismic surveys.

Regions in which operations are difficult or expensive may be explored by a company under contract with a consortium of oil companies. Seismic surveys of the outer continental shelf, for example, are usually conducted this way. After the federal government announces the impending sale of offshore leases, but before the sale, several companies may pool their exploration efforts and share the results, later basing their bids on the estimated probability of finding commercial quantities of oil in particular blocks.

As he learns more about a region, the geologist narrows his search. He concentrates on particular formations and studies local structural features, sediment types, permeability trends, possible source beds, and significant clues about depositional environments. As the subsurface picture comes into focus, he refines his quantitative knowledge: the thickness of the reservoir formation, areas of maximum porosity and permeability, potential capacity for oil. The size of a potential reserve is calculated for comparison with the costs of recovery.

By constructing this detailed geologic picture, the petroleum geologist strives to select the drilling site where his company is most likely to find oil, as well as to estimate how much oil is likely to be found should an accumulation actually be present. And, as we shall see, his predictions become more precise and accurate as more wells are drilled.

GEOLOGIC SURVEYS

Surface Data

The petroleum geologist continues the narrowing-down process by observing the most direct geologic evidence available: the surface. The character of underground formations and structures can often be deduced from what appears at the surface.

Early drillers produced oil from shallow wells near seeps. Knowing that tilted surface outcrops often extended to great depths, they found more oil by drilling deeper and farther from seeps. Gradually they learned that large oil pools were likely to be found near the axis of an anticline. Anticlines became much sought after, and most of the world's known reserves were subsequently found in them.

Anticlines with surface expression are relatively easy to locate (fig. 4.1). The eroded outcrops dip in both directions away from a central axis. Similarly, the presence of fault traps can often be deduced by careful observation and mapping of outcrops.

Oil and gas seeps are obvious signs of a subsurface petroleum source. Often, however, seepage is so slow that it is not easy to

(Photo by John S. Shelton)

Figure 4.1. Outcropping anticline (From *Geology Illustrated* by John S. Shelton. Copyright ©1966 by W. H. Freeman and Company. All rights reserved).

see; bacteria and weathering may decompose the oil as it nears the surface, preventing visible accumulation but altering the chemistry of the soil or water. Chemical testing for traces of hydrocarbons can often uncover such hidden seeps, leading to further exploration. Oil fields have been found at sea by using *sniffers* (chemically sensitive instruments) to detect plumes of gas rising from the ocean floor.

It is easier to see regional geologic patterns from a great height than from the surface. Aerial photography and satellite photography provide new information, especially on remote and undeveloped regions, and offer new perspectives on even well-developed areas.

High-altitude aerial photographs, including those taken by satellites such as Landsat, are commonly taken by using filters and special films to record light of specific wavelengths, both visible and invisible. Photographs taken in infrared light are useful in determining vegetation and rock outcrop patterns (fig. 4.2). Subtle differences in these patterns may indicate the

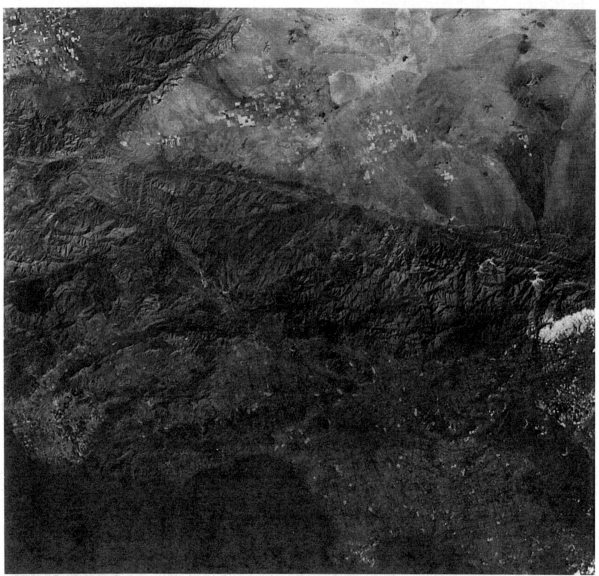

Figure 4.2. Landsat infrared photograph of Los Angeles area (Courtesy of United States Geological Survey EROS Data Center)

presence or absence of trace minerals in the soil—including hydrocarbons. Thus in certain instances it is possible to detect, from hundreds of miles up, hydrocarbon seeps whose existence was not suspected.

Visible-light aerial photos are often taken in stereo pairs that provide a three-dimensional image. Such photographs are useful in determining the dip and strike of outcrops, formation thickness, fault trends, and other structural characteristics. The most sophisticated application of aerial photography, however, is in discovering subtle relationships between seemingly unrelated surface features. Landscape features that a ground-based observer might overlook sometimes occur in distinct regional trends (*lineaments*) when seen from high above.

One relatively new type of remote sensing is *imaging radar*. High-frequency radio waves are bounced off surface features to one or both sides of an airplane or satellite, and the return echos recorded. This technique produces a light-and-shadow relief pattern showing topographic detail (fig. 4.3). The image is usually distorted and lacks information on rock types; but it is useful in searching unexplored areas for potential oil-trapping structures and for discerning large-scale terrain features at a glance.

Geophysical Surveying

The only sure way of finding out whether there is oil beneath the ground is to drill a well. But drilling is expensive, so the petroleum geologist resorts to methods that are more cost effective and provide data on larger areas. In doing so, he relies on a geophysicist—a specialist in using geophysical sensing techniques to find potential subsurface traps.

The principal geophysical surveying methods in use today are seismic, magnetic, and gravimetric. Seismic exploration provides the most detailed picture of subsurface geology but is by far the most expensive. In unexplored areas, the geophysicist conducts relatively inexpensive magnetic

Figure 4.3. Radar image of Paleocene rocks in Peru, from space shuttle *Challenger* (Courtesy of NASA)

and gravimetric surveys first to find the best areas in which to obtain the more definitive but costlier seismic profiles.

Magnetic surveys. The quickest and least expensive way to study gross subsurface geology over a broad area is magnetic surveying. A *magnetometer* is used to measure local variations in the strength of the earth's magnetic field and, indirectly, the thickness of sedimentary rock layers where oil might be found.

Most of the igneous or metamorphic basement rock that underlies sedimentary rock contains iron or titanium, metals that affect the earth's magnetic field. This field, although complex, can be likened to that of a bar magnet, around which the lines of magnetic force form smooth, evenly spaced curves (fig. 4.4). If a small piece of iron is placed within the bar magnet's field, it

becomes weakly magnetized, creating an *anomaly,* a distortion of the field in which the lines of force are concentrated and the field locally stronger.

Iron-bearing rocks in the crust concentrate the earth's magnetic field in much the same way. The degree of concentration depends not only upon the amount of iron or titanium present in the rock but also upon the depth of the rock. A body of igneous rock 1,000 feet down will affect a magnetometer more strongly than a similar mass 10,000 feet down. Thus areas of relatively low magnetic field strength are usually areas with the thickest sequence of nonmagnetic sedimentary rocks (fig. 4.5).

Once the magnetic readings have been plotted on a map, points of equal field strength are connected by contour lines (fig. 4.6). Such a map is roughly equivalent

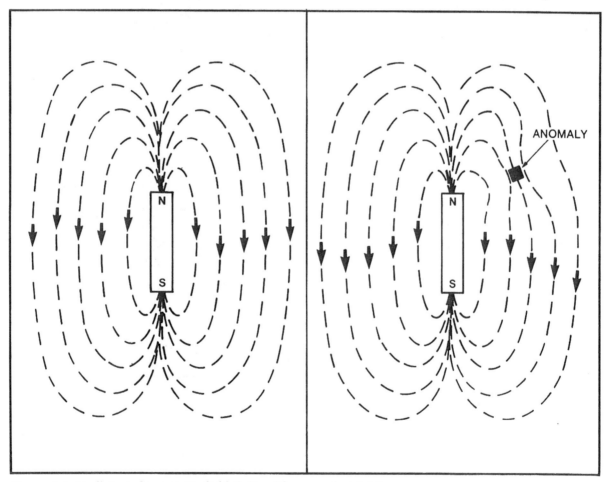

Figure 4.4. Undistorted magnetic field (*left*); with anomaly (*right*)

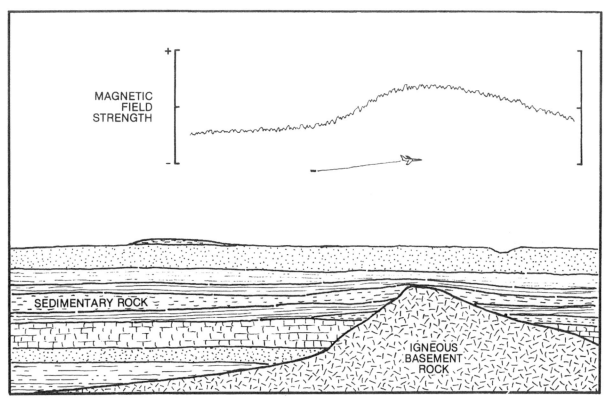

Figure 4.5. Magnetic surveying

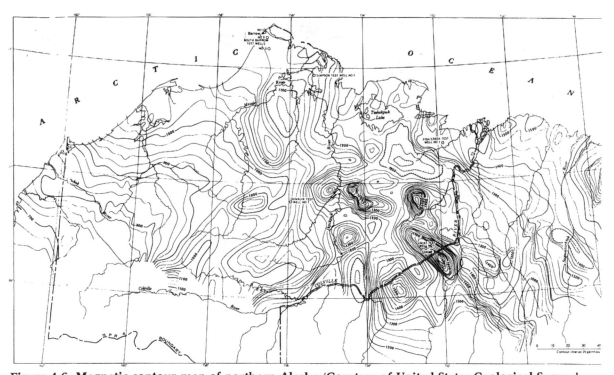

Figure 4.6. Magnetic contour map of northern Alaska (Courtesy of United States Geological Survey)

to a topographic map of the basement rock. It can be useful in locating grabens or other basins beneath miles of sedimentary rock, hidden columns of salt rising through heavier overlying strata, or basement rock associated with buried structural highs. However, magnetic contours are of little use in revealing details of structure or stratigraphy.

Most magnetometers used today are electronic instruments that can detect magnetic variations while on the move. A *fluxgate* instrument measures changes in field strength from point to point; the more accurate *proton magnetometer* measures absolute field strength at any point. Magnetometers can be used on the surface to survey large grids and produce detailed magnetic variation contour maps or towed behind ships or planes to survey large areas quickly.

Gravimetric surveys. Most igneous or metamorphic basement rock is denser than porous sedimentary rock and therefore exerts greater gravitational attraction. Locales with more basement rock and less sedimentary rock will therefore read higher on a gravity meter, or *gravimeter*. A gravimetric survey can reveal basement rock structure at costs comparable to those of magnetic surveys. In areas where deep sedimentary structure is masked by overlying igneous strata, such as the basalt of the Columbia Plateau in Washington, gravimetry is the preferred technique.

Although mechanically simple, a gravimeter can measure gravity anomalies as small as one billionth of the earth's surface gravity. Corrected for latitude, altitude, and other factors, these readings are plotted like contours of magnetic variation (fig. 4.7).

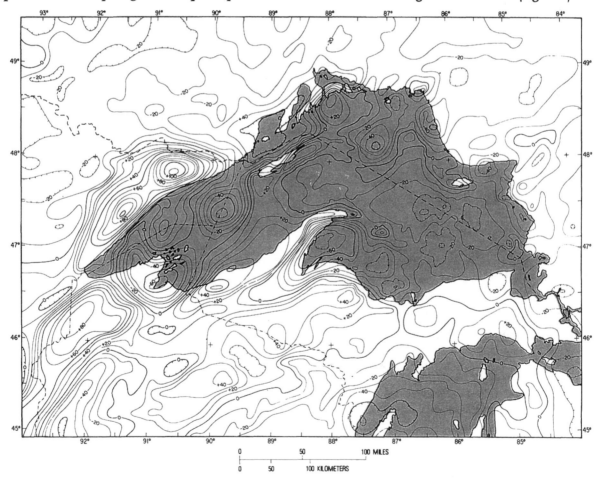

Figure 4.7. Gravity anomaly map of Lake Superior (Courtesy of United States Geological Survey)

Seismic surveys. The geophysical survey method that provides the most detailed picture is the seismic survey. Seismology is based on the transmission of sound waves by the rocks of the crust. Strong earthquakes create pressure waves that are transmitted through the entire earth and detected by seismographs on the other side. Seismic surveying, however, as employed by the petroleum geologist, makes use of artificially generated pulses. These pulses are much weaker but are focused on areas of stratigraphic interest.

The speed of seismic pressure waves through the crust varies directly with density and inversely with porosity. Through soil, the pulses travel as slowly as 1,000 feet per second, comparable to the speed of sound through air at sea level. Some metamorphic rocks, on the other hand, transmit seismic waves at 20,000 feet per second, or 4 miles per second.

Pressure waves induced at or near the surface, such as by an explosion, are transmitted downward and outward from the source until they encounter a *velocity boundary*, such as a boundary between rock types (fig. 4.8). Here they behave like light waves at an air-water interface: they change direction. Some are reflected toward the surface;

others are bent, or *refracted*, as they cross the boundary. Waves that strike the boundary at the *critical angle* (an angle determined by the acoustic properties of the rocks above and below the interface) are refracted parallel to the boundary and travel along it, radiating both downward and upward until their energy is dissipated some distance from the source. These refracted waves were the first type used by seismic oil prospectors ("doodlebuggers") to locate possible oil-bearing structures. The effectiveness of early refraction seismology was limited, however, by its inability to show details of structure or to resolve more than a few strata.

Like the underwater sound-pulse detection system known as *sonar*, seismic surveying involves measuring the time interval between the generation of a pressure pulse and the arrival at the surface of its reflection or refraction. If the velocity is known, this interval can be graphed in terms of distance traveled, or depth of the reflecting horizon.

The equipment used in obtaining a seismic profile includes a source of seismic pulses, a seismic detection system, a timing system, and a recording system. In a land-based operation, the initial pulse is often produced by detonating explosive charges

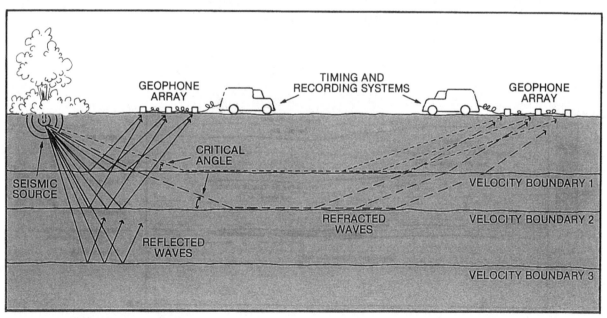

Figure 4.8. Seismic ray paths

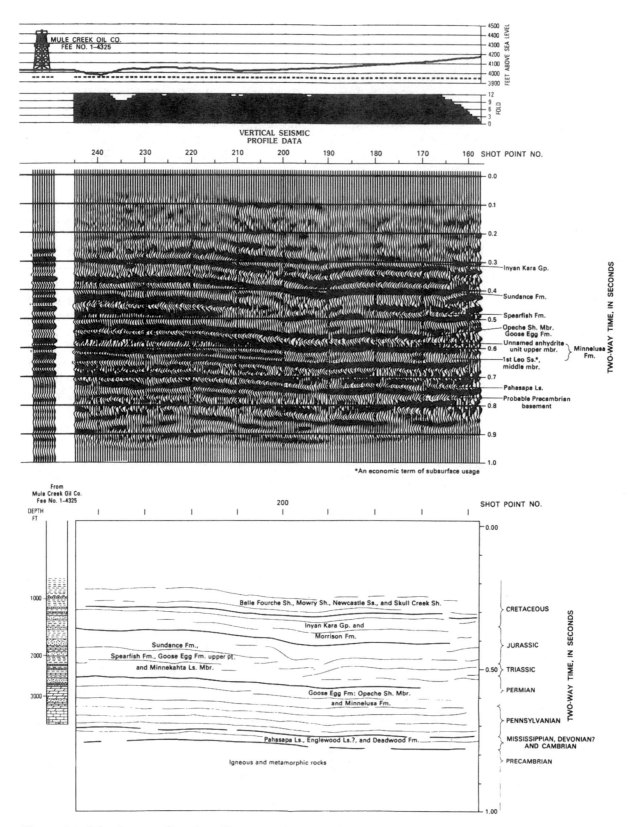

Figure 4.9. Seismic record section (Courtesy of Lee, et al., United States Geological Survey)

in one or more drilled holes up to 200 feet deep. Surface seismic generators include a truck-mounted detonation chamber using natural gas or other fuel to create a contained explosion; a truck-mounted, low-frequency mechanical vibrator; and a "thumper," a heavy slab dropped from a crane.

Reflected pulses are detected mechanically by an array of *geophones* and transmitted electronically via landline to timing and recording devices. The data are digitally recorded on magnetic tape and processed by computers into a *record section* that resembles a geologic cross section (fig. 4.9). A single seismic measurement onshore may provide a geologic profile along a line several thousand feet long.

Determining the character of each reflector in a seismic profile from a seismic record alone is rarely possible. Generally, however, the geologist knows enough about nearby formations to identify the seismic reflectors by their depth, thickness, and other characteristics. If not, a stratigraphic test hole may be drilled to ascertain rock types.

The raw data of a seismic record are produced by differences in the way various rock strata transmit sound. By itself, such a record does not give an accurate picture of subsurface strata because waves reflected from deeper layers are affected by their return trip through the overlying layers. Each reflected pulse is like light traveling through a series of lenses, reflecting off a mirror, and returning through the lenses before reaching the observer. Such lenses might make the mirror appear to be nearer or farther away than it actually is. The only way to accurately determine the distance to the mirror is to determine how each lens affects the returning light rays.

Similarly, the exploration geologist must adjust his results by determining how each overlying layer affects a reflected seismic pulse. The focusing effect of a layer depends on the velocity at which it transmits a seismic pulse. The geologist must determine this velocity for each layer. One way to do this is to place geophones in a well adjacent to a given stratum and generate a pulse in a nearby well or outcrop.

Because of a lack of velocity data, seismic surveys failed to reveal the existence of the Kelly-Snyder field, one of America's largest, until several wells had been drilled completely through it. Seismic reflections from the deep Ellenburger formation, a familiar West Texas oil-producing horizon, seemed to show a structure in which oil could accumulate. The reflections, however, had been distorted by the overlying Horseshoe Atoll reef limestone, whose great thickness was unsuspected. Only after dry holes had been drilled in the Ellenburger did geologists realize that the drillers had drilled through the huge Kelly-Snyder pool in the Canyon formation. Figure 4.10 shows a seismic section, adjusted with velocity data, of this limestone reef.

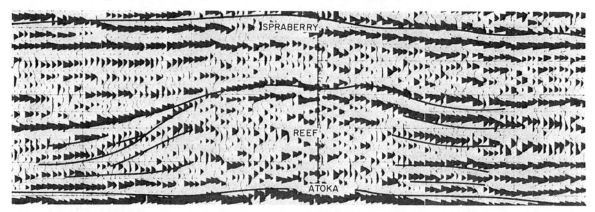

Figure 4.10. Seismic record section of Kelly-Snyder oil field, Texas (Courtesy of Galloway, et al., Bureau of Economic Geology, The University of Texas at Austin)

Recent developments in seismic exploration techniques include using powerful computers and special source and geophone arrays to paint a detailed three-dimensional picture of the survey area. Improved equipment and data processing can also enhance detail, in some cases enabling the geologist to identify an oil-gas or water-gas interface at considerable depth. Under favorable conditions, the velocity and density characteristics of strata above and below a reflection interface can be measured by seismic techniques, giving information on lithology, porosity, and fluid content.

Although some oil companies run their own seismic crews, most surveys are conducted by service companies that specialize in seismic techniques. Sometimes the seismic survey company will line up several subscribers in advance and conduct a group shoot. The service company will commonly survey a large area in a grid pattern; oil com-panies participating in the group shoot then obtain the data more cheaply than they could if they ran the survey themselves, but their competitors have access to the same data. Other surveys, including expensive high-resolution surveys, are usually done under contract for a company needing accurate information over a more localized prospect area.

Seismic exploration offshore is easier and faster than that on land because the source and geophones can be towed behind a ship across a grid covering hundreds of square miles (fig. 4.11). To avoid bubbles and other noise factors characteristic of underwater explosions, offshore techniques usually include using a spark chamber or other device to produce the pulse. Marine refraction surveys use geophones planted on the ocean floor to detect pulses from sources up to 50 miles away.

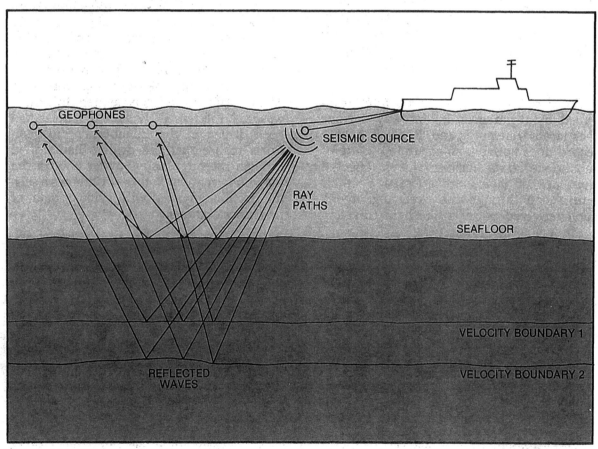

Figure 4.11. Offshore seismic surveying

Stratigraphic Testing

Geophysical surveys provide information on structural thicknesses and trends but tell the geologist little about the character of the formations. To find out what type of rock is represented by each seismic horizon, as well as important information such as porosity, permeability, and formation fluids present, the geologist must collect samples that he can examine directly, and he must perform tests on the rocks in place beneath the surface. These tasks are accomplished by means of a *stratigraphic (strat) test*.

The only way to get at the rock in a subsurface formation is to drill a hole. A strat test is a borehole that is drilled expressly to gather data on rock types and sequences. Some strat tests are drilled merely to locate or confirm the presence of known formations; others are conducted where the subsurface is totally unknown.

Strat test holes are usually smaller in diameter than wells drilled to locate or recover petroleum. They can be drilled with smaller rigs and fewer personnel and are therefore less expensive. However, development wells often contribute the same types of information.

As the strat test (or exploratory or development well) is being drilled, the geologist examines and identifies the first samples that become available: *cuttings*, the fragments of formation rock that are gouged out by the bit. Identifying cuttings does not provide an exact record of stratigraphic sequence because they become thoroughly mixed in the mud returning to the surface, and each sample taken from the shale shaker contains rock from different formations. But the appearance of a new rock type is an indication that a geological boundary has been crossed.

For greater precision and detailed analysis, the geologist needs a more coherent sample than the fragments produced by ordinary drilling. A special bit can be used to obtain a *core sample*—a solid cylinder or plug of rock from the formation being drilled (fig. 4.12). After the hole has been drilled,

Figure 4.12. Core sample (Photo by T. Gregston)

the geologist can obtain a sample of a particular formation from the wall of the hole. An explosive tool is lowered to a measured depth, and a hollow, retrievable bullet is fired into the formation to remove a *sidewall core sample*.

Cores can be analyzed in ways cuttings cannot. Because the sample is structurally unchanged, its porosity and permeability can be readily evaluated in the laboratory. Larger fossils can be identified and other significant details examined to determine the environmental conditions under which the sediment was deposited. The quantity of hydrocarbons present can more easily be determined in a core than in a sample of fragments that has been dispersed in the drilling fluid.

Cores may also be collected while drilling an exploratory or development well, but only when there is a special need for it, because coring is slower than ordinary drilling and drilling time on larger rigs is costly. Cores from an oilwell are often more useful than strat test cores because they can be larger in diameter.

Some formation characteristics are better analyzed in place than in a sample. Logging instruments are run into the hole on wireline to measure chemical and physical properties. The readings are transmitted to the surface via electrical wireline and are recorded digitally for computer processing and on strip chart recorders for visual analysis. Because the procedure is both inexpensive and richly rewarding in data, most wells are logged.

DOCUMENTING THE PROSPECT

The subject of the petroleum geologist's investigations is mostly hidden from view. In order to analyze the data and arrive at conclusions about the hydrocarbon potential in the rocks beneath the surface of a prospect, the geologist must create a picture of the subsurface that he can manipulate in his mind. And in order to communicate this knowledge to others, he must convert the data into usable forms: graphic representations in one, two, and three dimensions.

Geologic Mapping

The information that a geologist obtains about a prospect can be thought of as pieces of a three-dimensional puzzle. Some of these pieces have only one dimension, such as a core sample or a log showing a vertical sequence of sedimentary rock layers. Others are two-dimensional, containing information that can be represented on a map. But the sum of this knowledge is a three-dimensional array.

Sometimes the geologist may construct an actual three-dimensional geological model of a prospect. A typical peg model, for instance, is a set of sticks or pegs arranged to represent drilled wells or strat tests, marked to show the stratigraphic sequence at each site, and connected together with strings that represent upper and lower formation boundaries between sites. As can be imagined, a peg model is an elaborate and cumbersome device that is not easily portable. A more portable, but in some ways less detailed, model can be made of clear plastic.

Since three-dimensional geologic models are inconvenient to carry around, the geologist usually presents portions of the information in the two-dimensional form of maps and sections or sometimes as diagrams showing what a three-dimensional model might look like. These representations are constructed by using various data, such as formation thickness, rock type, porosity, the contours on an unconformity, or the ratios of particle sizes within a formation. By concentrating on a particular type of data, the geologist can exhibit a geologic fact that he deems significant without interference from other information. He saves the viewer the task of sorting out the important facts from the mass of data available. However, for a complete understanding of the prospect, the geologist may use many different types of maps and sections showing all relevant aspects.

Block diagrams. An elementary three-dimensional picture of a part of the earth's crust can be represented in two dimensions,

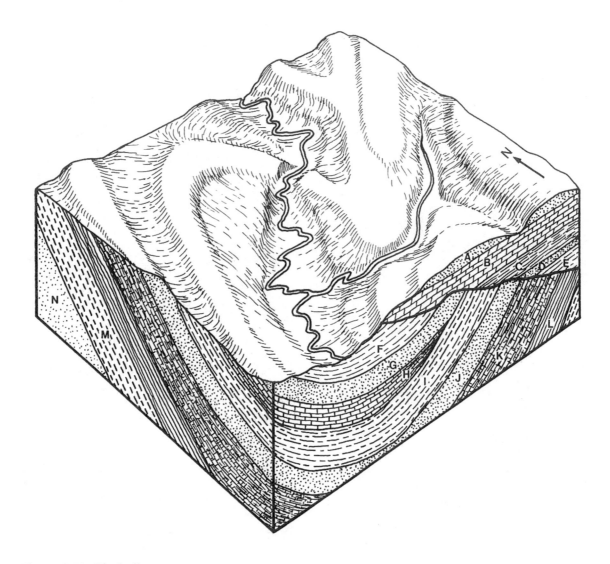

Figure 4.13. Block diagram

as it is in a *block diagram* (fig. 4.13). The block diagram is the most visually realistic type of presentation, especially if drawn in perspective. It shows a portion of the crust as it would appear if cut out of the earth like a serving of cornbread. The block diagram in figure 4.13 combines two vertical sections at right angles to each other (the faces of the block) and a perspective view of surface topography.

Topographic maps. If the viewer could somehow move his viewpoint directly above the "surface" of the block diagram,

what he would see could be made into a map. Maps show horizontally configured data. Most people are familiar with maps that show roads and towns and other surface features; many have seen topographic maps, such as those produced by the U.S. Geological Survey, showing height contours—lines connecting points at the same elevation above sea level. Topographic contours show the hiker where the mountains and the valleys are, where the slope is steep and where the hiking is easy.

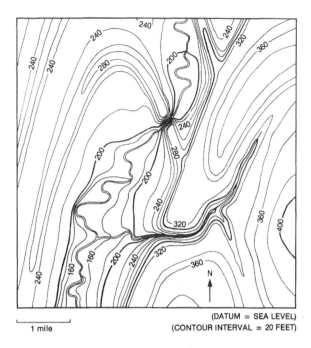

(DATUM = SEA LEVEL)
(CONTOUR INTERVAL = 20 FEET)

1 mile

Figure 4.14. Topographic contour map

Figure 4.14 is a *topographic map* of the area shown in the block diagram of figure 4.13. The elevation ranges from less than 160 feet above sea level at the southern end of the stream to more than 400 feet on the

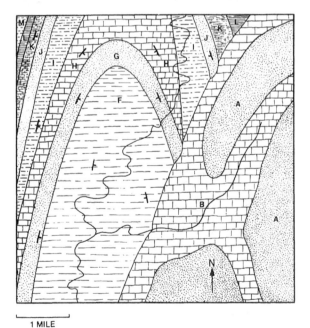

1 MILE

Figure 4.15. Geologic outcrop map

hill at the eastern edge of the map. The contour lines are widely spaced in the flat-floored valleys but very close together just above the center of the map, where the stream has cut a short, steep-walled canyon through the ridge.

Surface contours often hint strongly at what lies beneath. To the field geologist, the bow-shaped ridge is a strong clue of a possible folded structure, either an anticline plunging toward the north or a syncline plunging southward.

Geologic maps. A *geologic,* or *outcrop, map* shows the rock types at the surface or just beneath the soil. One of the most prominent features of the geologic map shown in figure 4.15 is the series of bow-shaped outcrops in its western half. Comparison with the topographic map (fig. 4.14) reveals that the bow-shaped ridge corresponds to the sandstone outcrop *G*, which is apparently more resistant to weathering and erosion than the adjacent shale *(F)* and limestone *(H)*. The plateau on the right is a horizontal sheet of limestone overlain by sandstone.

In this hypothetical landscape, field studies of the outcrops reveal that the rock layers *F* through *L* in the western half dip toward the axis of the valley. Closer examination of the formations and their fossil assemblages indicates that the youngest rocks occur near the axis and that the rocks become progressively older away from the axis. This evidence strongly supports the case for the southward-plunging syncline.

Cross sections. Any vertical face of the block diagram, viewed straight on, is a *cross section* (fig. 4.16). (The term is usually reserved for a vertical section but is occasionally applied to sections in other planes.) A cross section represents a portion of the crust as though it were a layer cake sliced by a huge knife, revealing the layers inside. Cross sections are usually drawn with the vertical dimension exaggerated as much as ten or even a hundred times. A cross section drawn to scale would be useless to show a formation 10 feet thick and 10 miles wide. In viewing the section, however, keep in mind that this vertical exaggeration distorts the shape of the formation, making dips

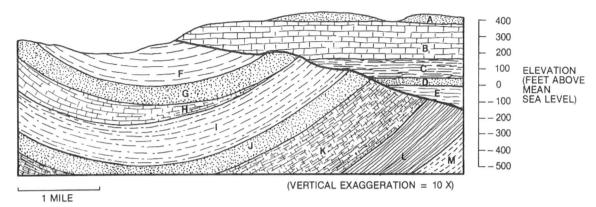

Figure 4.16. Cross section

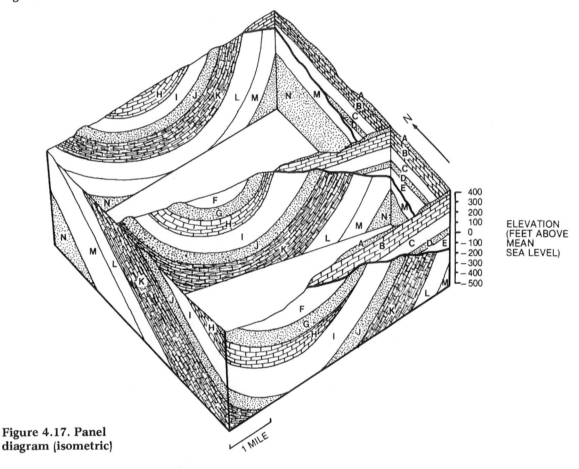

Figure 4.17. Panel
diagram (isometric)

look steeper than they actually are and dipping or plunging structures look thinner
than horizontal ones.

Most cross sections show both structural
and stratigraphic features together, with
perhaps more emphasis on one than on the
other. Figure 4.16 shows clearly the angular
unconformity where formations *B* through

E lap out on the former erosional surfaces of
formations *F* through *L*.

Panel diagrams. A series of intersecting
cross sections joined together and viewed
obliquely from above would resemble a
panel diagram (also called a *fence diagram*).
The panel diagram shown in figure 4.17 is
drawn isometrically – that is, without

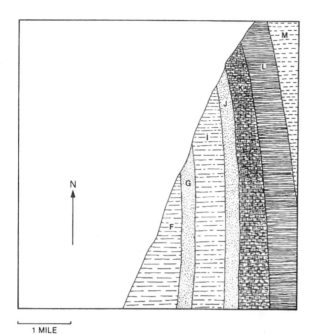

1 MILE

Figure 4.18. Paleogeologic map showing formations beneath unconformity

perspective—but otherwise resembles a hollowed-out block diagram. A panel diagram is useful in showing how formation structure and stratigraphic thickness vary both horizontally and vertically.

Paleogeologic maps. The geologist uses many different kinds of maps to represent what lies beneath the surface. If the geologist were able to remove all the strata above the unconformity in the hypothetical example, he would be able to see the surface of the land as it existed at some time in the past. A *paleogeologic*, or *subcrop, map* shows the rock types that constitute the buried landscape (fig. 4.18). The geologist can use such a map to locate likely reservoir beds, such as outcrops of sandstone or other elevated porous rocks.

Subgeologic maps. The bottom, as opposed to the top, of an unconformity is portrayed by a *subgeologic*, or *worm's-eye, map* (fig. 4.19). This type of map shows the formations that lap out on the buried landscape, as seen from above. It is the "footprint" of the formations overlying the buried erosional surface. A subgeologic map is useful in tracing the depositional histories of the rocks above an unconformity.

Structure contour maps. Contour lines can be used to show the shapes of hidden surfaces. One of the most useful contour maps is the *structure contour map* (fig. 4.20). To gain a clear idea of the structure of the

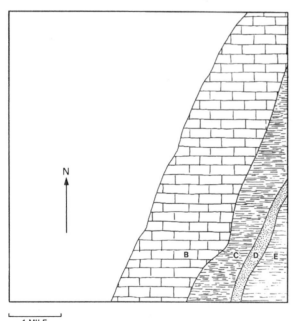

1 MILE

Figure 4.19. Subgeologic map showing rock types immediately overlying unconformity

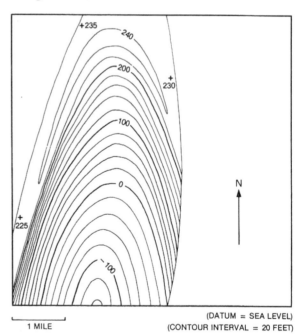

1 MILE

(DATUM = SEA LEVEL)
(CONTOUR INTERVAL = 20 FEET)

Figure 4.20. Structure contour map of top of horizon *H* (limestone)

limestone formation *H* (fig. 4.13), the geologist constructs a map that shows the elevations of the top of this layer. The principle is the same as that used in the topographic contour map of the surface.

Structure contour maps are important in locating subsurface structural highs, which often do not correspond with surface topography. If the axis of an outcropping anticline is tilted, for instance, the high points of deeper members of the folded series will be displaced laterally from those of the surface strata. The petroleum geologist is usually more concerned with the conformation of deeper layers, where petroleum migrates and accumulates, than with surface topography.

Isopach maps. Different types of maps can be used to portray different properties of the same horizon. Having determined the thickness of horizon *H* at a number of points, the geologist can draw a set of contour lines connecting points of equal thickness—an *isopach map* (fig. 4.21). Limestone *H*, which pinches out to the east beneath sandstone *G* (fig. 4.16), varies in thickness from zero at its edges to more than 160 feet at the southwest corner of the map.

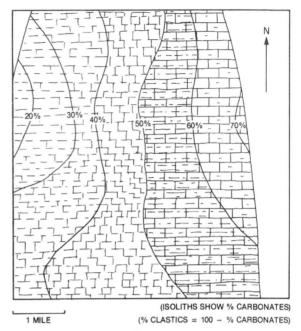

(ISOLITHS SHOW % CARBONATES)

1 MILE   (% CLASTICS = 100 − % CARBONATES)

Figure 4.22. Lithofacies map of horizon *K*

Lithofacies maps. If the geologist wishes to show how the character of the rock varies horizontally within a formation, he can construct a *lithofacies map,* one of many different types of facies maps. Suppose that horizon *K* varies continuously from approximately 20% carbonate and 80% clastic rock at its western edge to 70% carbonate and 30% clastic on the east. Contours can be drawn connecting points of equal carbonate percentage (or alternatively, equal ratios of carbonates to clastics), showing the lateral variations in rock type (fig. 4.22). Geologists use similar maps to show lateral variations in porosity, permeability, and other factors. A facies map showing variations in the occurrence of fossil types is a *biofacies map.*

Computer graphics. Powerful computers now coming into common use enable the geologist to manipulate three-dimensional data graphically on a video screen as though turning a solid model in his hand. Using sophisticated programming, he can create from raw stratigraphic data an image of the subsurface geology in which horizons can be blanked out or hidden details highlighted by a keystroke. Similarly, computer-linked automatic plotters can

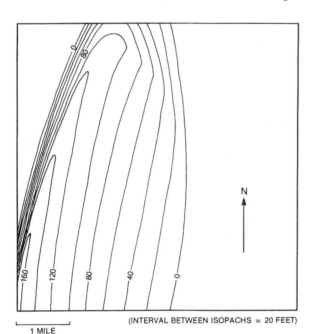

(INTERVAL BETWEEN ISOPACHS = 20 FEET)

1 MILE

Figure 4.21. Isopach map of horizon *H* (limestone)

Figure 4.23. Computer-generated isometric diagram of Horseshoe Atoll, Texas (From *Atlas of Major Texas Oil Reservoirs* by W. E. Galloway, et al., 1983, Bureau of Economic Geology, The University of Texas at Austin)

draw different views of the same geologic feature. Figure 4.23 shows a computer-generated isometric diagram (beneath a generalized structure contour map, also seen isometrically) of the Horseshoe Atoll óf West Texas (fig. 2.26). Oil fields in the higher parts of the reef are indicated by shading. Figure 4.24 is a computer-generated isometric diagram of the Hainesville Dome, also shown in the structure contour map of figure 3.15.

What Makes a Good Prospect?

The decision to drill an exploratory well rests on many factors, only some of which are related to geology. Legal and physical access to drilling sites is one consideration; others include the cost of drilling, the cost of potential production, and market prices. Nevertheless, the most important part of this evaluation is supplied by the geologist.

The geologist's information bears directly on the other factors that go into the final decision to drill: access, production costs, and especially the potential of discovering a commercially recoverable resource. In his evaluation of the prospect, the geologist must address all the factors that affect the company's chances of realizing a profit by drilling.

Location. Geographic considerations figure heavily in the potential value of the prospect. Legal access is a prime factor. Can

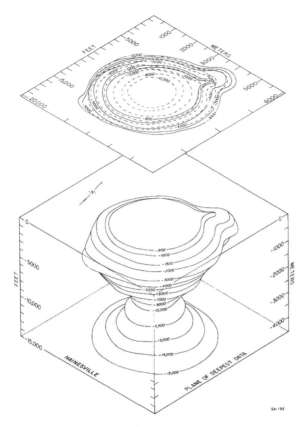

Figure 4.24. Computer-generated isometric diagram of Hainesville salt dome, Texas (From R. I. 140 by Martin Jackson and Steve Seni, Bureau of Economic Geology, The University of Texas at Austin)

a lease be obtained? Most private land-owners and mineral-interest owners are willing to deal for a share of the resource; but if the resource is located beneath public land, the oil company may have to wait for the area to be opened to competitive bidding. Some federal lands on which cultural or aesthetic values have been ruled of overriding interest are permanently off limits to mineral exploration and development.

Physical access is another aspect of the value of a prospect. Is it located beneath the Mississippi River, Pike's Peak, or downtown Houston? Such prospects may be accessible by directional drilling, but this technique may so add to the cost of exploration and recovery as to render it impractical. Drilling an underwater location requires a much greater investment in equipment and sup-

plies; because of the additional cost, reservoirs deemed feasible to be drilled offshore must contain more recoverable reserves than if they were located onshore.

Drilling deep prospects requires heavier equipment and takes longer than drilling shallow wells. Faced with a choice of prospects at various depths, the oil company will choose to drill the shallowest, all else being equal, because it requires the smallest investment of time, equipment, and money. Offshore prospects must also take into account the depth of water at the drilling site.

Another factor that affects physical accessibility is the hardness or drillability of the overlying rock. A deep prospect beneath easily drilled rock may be more attractive than a shallower prospect located in crooked-hole country where bits must be replaced often and the likelihood of a fishing job is high.

Presence of source beds. Part of the geologist's task is to ascertain whether there is a potential source of hydrocarbons for the prospect. Shales rich in organic content may lie directly beneath a likely reservoir formation, or in contact with it many miles downdip; in either case, there is a route by which petroleum might have migrated into a reservoir (fig. 4.25). In some cases, the source rock may overlie the reservoir rock or be in lateral contact with it.

The organic content of the source beds may be quite apparent in an outcrop or a core sample. Even if the rock is low in organic residue, it may once have contained a considerable amount of organic debris that has since been converted to oil and squeezed out. Oil can form, migrate, and accumulate in a trap in less than a million years. Thus, although the age of the source rock is a factor, even relatively young potential source beds cannot be discounted.

The geologic history of the source rock has a significance beyond its age alone. To understand its potential fully, the geologist must reconstruct the events that led to the present composition and configuration of rock containing organic material. The conditions under which it was deposited – its depositional environment – had to include a

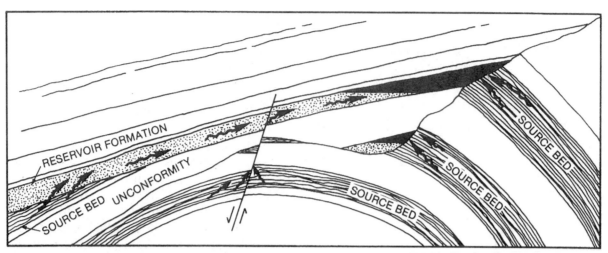

Figure 4.25. Source beds and migration routes

biotic community capable of supplying a quantity of organic detritus, as well as an anaerobic zone where detritus could accumulate without being oxidized.

Source rock must be buried deep enough to raise the temperature to that required for the formation of hydrocarbons from organic debris, but not so deep that excessive heat carbonizes the hydrocarbons. The set of physical conditions under which oil and gas can form—the oil window—is shown in figure 3.3. By reconstructing the geologic history of the prospect, the geologist can include in his report the likelihood that the source rock went through the oil window at some time in the past.

Presence of a trap. Rock of low permeability must be present to stop the migration of oil or gas. It must be strategically positioned both in space and in time. The seal must be in place before petroleum begins to reach the trapping structure, or it will continue to migrate. Oil may seep to the surface from an outcropping anticline, and later the erosional surface may be overlain with shale. If depleted of mobile hydrocarbons, the original source bed can no longer serve as a source of oil and gas. Oil may enter the trap from the younger, overlying source beds (which may also act as a seal) but not from the older source beds beneath (fig. 4.26).

A fault containing a low-permeability gouge or clay seal can stop hydrocarbon migration through otherwise permeable strata. However, if the rock affected is brittle, it may instead form a brecciated zone that is highly permeable to oil. Of course, a brecciated zone that does not reach the surface may become either a conduit to a higher reservoir structure or a trap that accumulates hydrocarbons (fig. 3.25). Most trapping faults work because an impermeable formation has been brought into lateral contact with a reservoir formation in such a way as to stop updip migration.

In many stratigraphic traps, updip facies changes within the reservoir formation stop hydrocarbon migration. Even in this case, though, the overlying beds must be impermeable. Across an unconformity, as across a fault, the truncated reservoir bed must abut an impermeable formation.

Properties of the producing zone. Generally, the presence of a body of rock that is porous and permeable enough to produce petroleum at commercial rates is known early in the geologic investigation. Much of the task of building a prospect consists of gathering information about associated factors. However, once it is determined that the chances of finding oil are high, the reservoir itself must be described in greater detail.

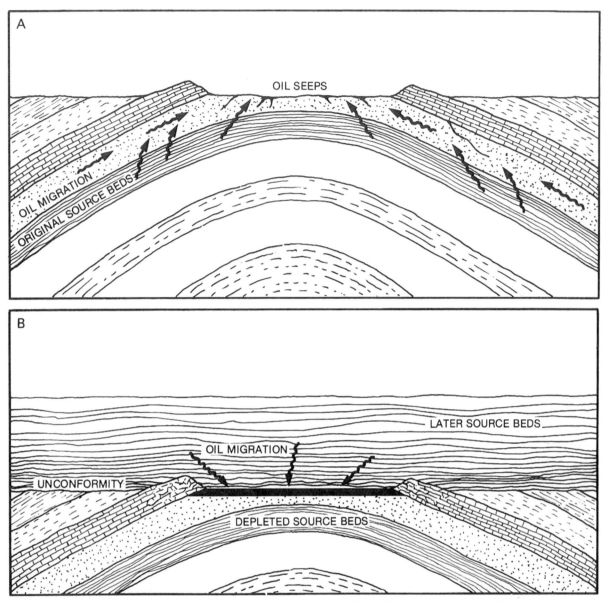

Figure 4.26. Anticlines: *A*, outcropping, with seeps; *B*, buried, with recent oil accumulation.

Reservoir porosity and permeability are essential data. With notable exceptions, formations with less than 10% porosity are not considered suitable reservoir rock. Even higher porosities, 20% or more, may not be considered worth exploring if other factors are adverse. Effective porosity, the percentage of the rock's total volume that is occupied by interconnected pores, is more important than total porosity because oil cannot be produced from isolated pores.

Permeability is a measure of how readily the reservoir can be made to yield its hydrocarbons. Even a large pool of oil loses much of its economic attractiveness if it can only be eked out by long-term pumping or other energy-intensive means. The viscosity of the oil is important in this connection, too. Permeability is measured in terms of the movement of a particular fluid; a rock through which a light hydrocarbon might readily flow may act as a seal or trap for a viscous crude. Huge reserves of tarlike viscous oil lie untapped around the globe for want of an economically practical recovery method.

Gas affects the oil in a reservoir in several ways. Dissolved gas reduces the oil's viscosity, making it easier to produce. During production, however, reduced pressure may allow the gas to come out of solution and form or augment a gas cap. An expanding gas cap helps drive the oil to the surface. Usually, however, the pressure of the gas cap is exhausted long before all the oil has been recovered.

Reservoirs below about 20,000 feet, as well as those whose geologic history includes deep burial followed by partial uplift, produce mostly gas, because oil that was trapped in them before they were deeply buried exceeded the temperature limits of the oil window and was converted to gas. The deeper formations in the Anadarko Basin are examples of gas-producing structures that probably once contained oil much like that produced from shallower wells nearby.

Except in rare oil-wet reservoirs, water fills the smallest pore spaces and coats the grains of rock, decreasing the effective permeability for oil. The proportion of water to oil in a reservoir is an important economic consideration; the higher the percentage of water, the less oil is present and the more difficult it will be to recover.

The movement of water is believed to be a major factor in the migration and accumulation of oil. Water at the oil-water interface acts as a lower reservoir seal. If the water can continue to flow in, its pressure drives oil into the well. Water is much more efficient at pushing oil out of a reservoir than gas; while gas-drive reservoirs may produce only 10% to 20% of the oil present, water drives commonly produce 40% to 50%.

Reservoir dimensions. Not least among the geologist's responsibilities is estimating the amount of the possible reserve in the prospect. This volume is a function of the thickness of the reservoir bed, its area, its closure (the vertical distance from the top of the reservoir to its lowest open contour), and its effective porosity. Reservoir volume represents the maximum amount of petroleum that the reservoir can produce—a figure that must be adjusted after determining reservoir pressure, drive, and the types, ratios, and properties of the fluids.

———

The geologist cannot guarantee that oil will be found in a particular location. He can at best predict the presence of a geologic trap capable of accumulating hydrocarbons and provide an educated guess about the probability of its containing oil. About one out of three suspected traps is found to exist, and perhaps one out of four traps is found to contain recoverable petroleum. Only when the well is finally drilled and oil comes into the wellbore can anyone be certain of hitting pay dirt. The geologist's task boils down to making it as easy as possible to assess the risks of drilling an exploratory well.

5

ECONOMICS

When the ultimate decision of whether or not to drill is made, other data are evaluated along with all of the preliminary geophysical and geological evidence. Many of these added considerations are financial and practical ones. A modern geologist has to be acquainted with the factors influencing wellhead and product pricing, transportation costs, fluctuations in supply and demand, associated political situations, and regulation and taxation, as well as the current costs of drilling.

All of these factors are balanced and weighed, often by complex computer programs. Much of this type of evaluation is done at the managerial level of the operating company; however, since the geologist is often concurrently evaluating physical evidence, the two types of data become closely related. Therefore, some economic and legal knowledge on the part of the geologist is required.

In a large oil company, much expertise is available for making the ultimate decision on what to do with leased lands. Often present at a decision meeting are the heads of staffs from such departments as exploration, production, engineering, and economic analysis. Probably a high executive, the chief landman, and a staff geologist will be there (fig. 5.1). In a small independent company, one individual may

have to balance all these roles, requiring a knowledge of engineering, mathematics, statistics, probabilities, and geology and geophysics.

Sometimes competition concerning where to invest a limited amount of funds exists among the departments of an oil company. For instance, the geologists might want to drill a wildcat in an area where they have recently invested their time and which they have decided has promise. On the other hand, the production engineers might want the same funds to follow up on the development of a field where they've predicted fairly certain further producible reserves. A management decision would therefore be necessary, and it would include consideration of the differences in capital requirements of the two ventures. The timing of cash flow concerning early or late payout on investment can be crucial, depending on the company's cash flow status at that particular time.

A petroleum geologist has to have the knowledge to weigh such decisions even if computers, more and more, are doing the actual calculations necessary to the economic analysis so prominent in petroleum evaluations today. In order to question a machine, the geologist must know what to ask and have a basic understanding of how factors are being weighed. The computer

Figure 5.1. Geologist explaining decision tree to a company meeting for exploration planning

becomes a valuable tool for saving time and personnel in tedious mathematical operations, but the competent geologist has an understanding of what is being weighed and how. The computer becomes especially valuable in exploration activity by performing "what if" sensitivity analyses, with multiple criteria allowed.

Frequently, the decision to be made will be whether the oil company holding leased lands wants to do the drilling or wants to *farm out*. To farm out is to assign part of the leasehold interest to a third party, usually another operator. The arrangement is conditional upon the third party's drilling a well within the expiration date of the primary term of the lease. By farming out parts of the lease (including the proportionate share of the drilling costs), the oil company can reduce some of its risk. The potential of gain is also lessened, of course, since the profit, if

any, is shared with the third party taking the farmout.

The decision-making process will differ greatly according to what type of drilling venture is being contemplated. At the extremes lie drilling a rank wildcat in a virgin basin and developing a field with proved reserves. Industry-wide usage of the terms *proved* and *unproved* reserves varies; therefore, the terms can be understood only in a general sense. The capability of estimating the quantity of hydrocarbons in the ground has been improved enormously through technological advances over the years. However, the accuracy of those estimates cannot be guaranteed. In general, the term *proved reserves* refers to the estimated volume of crude in a field that has been developed sufficiently to make reasonably accurate predictions possible. Obviously, making a decision with these kinds of data eliminates a good deal of the risk that is involved in a strictly wildcat venture.

When exploration activity begins on a prospect, its ultimate goal is the development of a field. A *field,* in common terminology, is the general area underlain by one or more reservoirs formed by a common structural or stratigraphic feature.[1] The Railroad Commission of Texas uses the term to refer also to a single reservoir. The ideal game plan for the explorationist is to drill a wildcat, or exploratory, well first. A *wildcat* is a well drilled in an area where no oil or gas production exists. If the wildcat shows promise of enough oil or gas to indicate commercial success in the field as a whole, the wildcat becomes what is termed a *discovery* well. Several exploratory wells may be drilled on the prospect.

If the discovery well shows enough promise, then the field will be developed with *development* wells in a plan that considers both financial reward and the legal status of the field. Explorationists know that the odds favor one or more dry holes along with the producible development wells. A *dry hole* is any well that does not produce oil or gas in commercial quantities. The cost of dry holes must be calculated in all prefiguring if the cost analysis for the field is to be as accurate as possible.

The distinction between exploratory and development wells is even sharper in drilling offshore. Exploratory wells are drilled from mobile rigs—submersibles, semisubmersibles, jackup rigs, or drill ships. Leasing one of these drilling facilities is cheaper than building fixed platforms over and over on the ocean floor. When enough information concerning field size and reservoir drive has been gathered from exploratory wells, the wells are plugged and abandoned. At the optimum location, a fixed platform is built, often at great expense, and the development wells are drilled and produced from this platform. A platform can be designed for the drilling of a number of wellbores with a movable derrick that can

be skidded from one wellhead location hole to another. Different areas of the reservoir can then be penetrated by directional drilling (fig. 5.2).

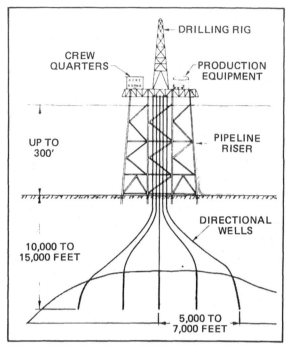

Figure 5.2. Directional drilling

SUMMARY OF AN ECONOMIC ANALYSIS

What then is the process that follows when a geologist has sufficiently promising information on a geological formation to recommend to his company that it might consider drilling? Money will probably already have been spent for seismic (or other geophysical) data and leasing activities. Often, these previous expenses are excluded from economic evaluation because the decision point is *now* in year zero. However, the interpretations of the geological data and the leasehold status are reviewed carefully by management and the economic advisers helping to make the decision. Correlation (the use of subsurface information from wells in nearby or similar formations) is an extremely important activity at this time. The size of the reservoir and the volume of its estimated recovery

---

1. Robert E. Megill, *An Introduction to Exploration Economics,* 2d ed. (Tulsa: PennWell Books, 1979), 14.

has probably been predicted by the geologist, who will be asked to defend and explain his opinion. Several cases, or extremes, of the geologist's prediction, from the most favorable to the most likely to the least favorable, will be charted.

At this point, an economic analyst will probably convert the predicted volumes of reserves into financial terms according to his understanding of future pricing, political events, regulations, and transportation charges. He will predict the cost of the first well, the cost and number of dry holes, and the cost and number of development wells if the exploratory well is a success. Thus, a *cash flow analysis*, or economic model of inflow and outflow, can be developed. The cash flow analysis may predict also the general plan to be used for the figuring of federal income taxing on that project.

After the cash flow analysis is studied, the company should perhaps be expected to have a fairly sharp picture of what to predict for the proposed venture. In fact, though, the high-risk/high-reward condition that is so indicative of petroleum exploration must still be dealt with. Since only about one in nine wildcat wells in untested areas finds evidence of hydrocarbons, and only one in forty leads to economic success, it is apparent that investment in drilling is fraught with risk. However, financial experts have developed methods of quantifying the various risks involved. The basic step in quantifying risk involves two variables: (1) the odds that no hydrocarbons at all will be found (the often quoted one-in-nine wildcat ratio), and (2) the extreme and the most likely cases concerning the volume of reserves if the wildcat is a success. In a process that will be described later, the *probability* of the various combinations of these outcomes is charted in a way that makes intelligent choices more likely and not so dependent on personal hunches. A *decision tree* can be drawn, demonstrating the predicted financial gain or loss of several possible outcomes for each possible action.

These then are some of the considerations of a typical meeting held to decide whether

or not to drill in a given area. Although many of the calculations and even interpretations involved are assisted by computer, the basic theories and conditions used in performing each of these stages of economic analysis must be understood. These stages of analysis concern availability of mineral rights, evaluation of estimated reserves, regulation considerations, taxation considerations, cash flow analysis, present value concept, and risk analysis.

AVAILABILITY OF MINERAL RIGHTS

Influential at this time is the leasing action that has been taken on the properties being contemplated. Leasing activity is often complex, particularly in the United States, which is the only country in the world where a significant number of mineral rights are not owned by the government. Often, a great deal of study must be made of the history of title to the rights. Although offshore minerals are owned by federal and state governments, the minerals below onshore land are frequently owned by individuals—farmers or other types of landowners. To make matters more complicated, the mineral rights are often owned by a different individual from the one owning the surface above them. Sometimes many individuals own percentages of the mineral rights. In order to contact the possibly large group of individual owners, most oil and gas companies employ one or more landmen.

A *landman* is a person who negotiates with landowners for land options, and with mineral holders for mineral rights leases. The landman contacts the landowner and mineral rights holder and offers on behalf of an oil company to lease the mineral rights for a certain period of time. The assumption behind this accepted practice is that the owner would have neither the expertise nor the capital to be the operator himself. If agreement is reached on terms, a lease stating those terms is drawn up.

A lease is a contract between the lessor (the mineral rights owner granting the lease) and the lessee (the oil or gas company seeking the lease). The lease stipulates a *royalty* payment to the lessor. A royalty is a negotiated percentage of gross income from a well. The remaining interest in the well – that of the lessee or oil company – is referred to as the *working interest*. For instance, if the landowner's royalty is $3/_{16}$, the working interest is $13/_{16}$. The lessee is responsible for all drilling and production costs and assumes the risk involved. The royalty owner pays only for taxes on his share if production occurs.

In the lease, the lessor will be paid an up-front fee, a *bonus*, usually figured on the basis of acreage. Also figured on the basis of acreage will be the amount of annual delay rental fees if the operating company does not obtain production within a stipulated amount of time. See figure 5.3 for an example of a lease.

Producers 88 (12-79) Revised
With 320 Acres Pooling Provision

POUND PRINTING & STATIONERY COMPANY
2325 Fannin, Houston, Texas 77002 (713) 659-3150

OIL, GAS AND MINERAL LEASE

THIS AGREEMENT made this _____ day of _____ 19___ between

_____

Lessor (whether one or more), whose address is: _____

_____ Zip Code _____

and_____

Lessee, (whether one or more), whose address is: _____

_____ Zip Code _____.

WITNESSETH:

1. Lessor in consideration of_____
Dollars ($_____), in hand paid, of the royalties herein provided, and of the agreements of Lessee herein contained, hereby grants, leases and lets exclusively unto Lessee for the purpose of investigating, exploring, prospecting, drilling and mining for and producing oil, gas and all other minerals, conducting exploration, geologic and geophysical surveys by seismograph, core test, gravity and magnetic methods, injecting gas, water and other fluids, and air into subsurface strata, laying pipe lines, building roads, tanks, power stations, telephone lines and other structures thereon and on, over and across lands owned or claimed by Lessor adjacent and contiguous thereto, to produce, save, take care of, treat, transport and

own said products, and housing its employees, the following described land in _____ County, Texas, to-wit:

This lease also covers and includes all land owned or claimed by Lessor adjacent or contiguous to the land particularly described above, whether the same be in said survey or surveys or in adjacent surveys, although not included within the boundaries of the land particularly described above. For all purposes of this lease, said land is estimated to comprise _____ acres, whether it actually comprises more or less.

2. Subject to the other provisions herein contained, this lease shall be for a term of _____ ( ) years from this date (called "primary term") and as long thereafter as oil, gas or other mineral is produced from said land or land with which said land is pooled hereunder.

3. The royalties to be paid by Lessee are:
(a) On oil, one-eighth of that produced and saved from said land, the same to be delivered at the well. If Lessor elects not to take delivery of the royalty oil, Lessee may from time to time sell the royalty oil in its possession, paying to Lessor therefor the net proceeds derived by Lessee from the sale of such royalty oil. Lessor's royalty interest in oil shall bear its proportionate part of the cost of treating the oil to render it marketable oil and, if there is no available pipeline, its proportionate part of the cost of all trucking charges.
(b) On gas, including all gases, liquid hydrocarbons and their respective constituent elements, casinghead gas or other gaseous substance, produced from said land and sold or used off the premises or for the extraction of gasoline or other product therefrom, the market value at the well on one-eighth of the gas so sold or used, provided that on gas sold at the well the royalty shall be one-eighth of the net proceeds derived from such sale. Lessor's royalty interest in gas, including all gases, liquid hydrocarbons and their respective constituent elements, casinghead gas or other gaseous substance, shall bear its proportionate part of the cost of all compressing, treating, dehydrating and transporting incurred in marketing the gas so sold at the wells.

Figure 5.3. Example of a Producers 88 lease form (Courtesy of Pound Printing)

(c) On all other minerals mined and marketed, one-tenth either in kind or value at the well or mine, at Lessee's election, except that on sulphur mined and marketed the royalty shall be fifty cents ($.50) per long ton.

(d) While there is a gas well on said land or on lands pooled therewith and if gas is not being sold or used off the premises for a period in excess of three full consecutive calendar months, and this lease is not then being maintained in force and effect under the other provisions hereof, Lessee shall tender or pay to Lessor annually at any time during the lease anniversary month of each year immediately succeeding any lease year in which a shut-in period occurred one-twelfth (1/12) of the sum of $1.00 per acre for the acreage then covered by this lease as shut-in royalty for each full calendar month in the preceding lease year that this lease was continued in force solely and exclusively by reason of the provisions of this paragraph. If such payment of shut-in royalty is so made or tendered by Lessee to Lessor, it shall be considered that this lease is producing gas in paying quantities and this lease shall not terminate, but remain in force and effect. The term "lease anniversary month" means that calendar month in which this lease is dated. The term "Lease year" means the calendar month in which the lease is dated, plus the eleven succeeding calendar months.

(e) If the price of any oil, gas, or other minerals produced hereunder is regulated by any governmental authority, the value of same for the purpose of computing the royalties hereunder shall not be in excess of the price permitted by such regulation. Should it ever be determined by any governmental authority, or any court of final jurisdiction, or otherwise, that the Lessee is required to make any refund on oil, gas, or other minerals produced or sold by Lessee hereunder, then the Lessor shall bear his proportionate part of the cost of any such refund to the extent that royalties paid to Lessor have exceeded the permitted price, plus any interest thereon ordered by the regulatory authority or court, or agreed to by Lessee. If Lessee advances funds to satisfy Lessor's proportionate part of such refund, Lessee shall be subrogated to the refund order or refund claim, with the right to enforce same for Lessor's proportionate contribution, and with the right to apply rentals and royalties accruing hereunder toward satisfying Lessor's refund obligations.

(f) Lessee shall have free use of oil, gas, coal, water from said land, except water from Lessor's wells, for all operations hereunder, and the royalty on oil, gas and coal shall be computed after deducting any so used.

4. Notwithstanding anything herein to the contrary, it is a condition of this lease that it shall not terminate upon any failure of the Lessee, for whatever reason, to make payments of any required shut-in royalty or rentals, either or both, herein provided for on or before the due date thereof unless and until: (1) Lessor notifies Lessee in writing by registered mail or certified mail, return receipt requested, of non-payment of the shut-in royalty or rentals; and (2) Thereafter Lessee fails to make payment of the shut-in royalty or rentals to Lessor within fifteen (15) days following Lessee's actual receipt of such written notice. Payment of shut-in royalty or rentals by Lessee to Lessor within fifteen (15) days following Lessee's actual receipt of said written notice from Lessor shall be deemed timely and sufficient to maintain this lease in force and effect. The provision of this paragraph are a part of the consideration for this lease, are contractual, and constitute a warranty from Lessor to Lessee. It is the desire and agreement of Lessor and Lessee to avoid forfeiture of this lease should Lessee fail to make payment of any required shut-in royalty or rental on or before the scheduled due dates thereof, and to afford Lessee an opportunity to make such payments within fifteen (15) days following actual receipt of written notice of non-payment from Lessor, thereby maintaining this lease in force. Such written notice from Lessor to Lessee shall state the full particulars concerning non-payment of shut-in royalty or rentals, identify the lease and land involved, the due date and amount claimed by Lessor, and Lessor's full name, current address and telephone number.

5. (a) Lessee, at its option, is hereby given the right and power to pool, unitize or combine the acreage covered by this lease or any portion thereof as to oil and gas, or either of them, with any other land covered by this lease, and/or with any other land, lease or leases in the immediate vicinity thereof to the extent hereinafter stipulated, when in Lessee's judgment it is necessary or advisable to do so in order properly to explore, or to develop and operate said leased premises in compliance with the spacing rules of the Railroad Commission of Texas, or other lawful authority, or when to do so would, in the judgment of Lessee, promote the conservation of oil and gas in and under and that may be produced from said premises. Units pooled for oil hereunder shall not substantially exceed 40 acres each in area, plus a tolerance of ten percent (10%) thereof, and units pooled for gas hereunder shall not substantially exceed in area 320 acres each plus a tolerance of ten percent (10%) thereof, provided that should governmental authority having jurisdiction prescribe or permit the creation of units larger than those specified, for the drilling or operation of a well at a regular location or for obtaining maximum allowable from any well to be drilled, drilling or already drilled, units thereafter created may conform substantially in size with those prescribed or permitted by government regulations.

(b) Lessee under the provisions hereof may pool or combine acreage covered by this lease or any portion thereof as above provided as to oil in any one or more strata and as to gas in any one or more strata. The units formed by pooling as to any stratum or strata need not conform in size or area with the unit or units into which the lease is pooled or combined as to any other stratum or strata, and oil units need not conform as to area with gas units. The pooling in one or more instances shall not exhaust the rights of the Lessee hereunder to pool this lease or portions thereof into other units. Upon execution by Lessee of an instrument describing and designating the pooled acreage as a pooled unit, said unit shall be effective as to all parties hereto, their heirs, successors, and assigns, irrespective of whether or not the unit is likewise effective as to all other owners of surface, mineral, royalty, or other rights in land included in such unit. Within a reasonable time following the execution of said instrument so designating the pooled unit, Lessee shall file said instrument for record in the appropriate records of the county in which the leased premises are situated. Any unit so formed may be re-formed, increased, decreased, or changed in configuration, at the election of Lessee, at any time and from time to time after the original forming thereof, and Lessee may vacate any unit formed by it hereunder by instrument in writing filed for record in said county at any time when there is no unitized substance being produced from such unit.

(c) Lessee may at its election exercise its pooling option before or after commencing operations for or completing an oil or gas well on the leased premises, and the pooled unit may include, but it is not required to include, land or leases upon which a well capable of producing oil or gas in paying quantities has theretofore been completed or upon which operations for the drilling of a well for oil or gas have theretofore been commenced. In the event of operations for drilling on or production of oil or gas from any part of a pooled unit which includes all or a portion of the land covered by this lease, regardless of whether such operations for drilling were commenced or such production was secured before or after the execution of this instrument or the instrument designating the pooled unit such operations shall be considered as operations for drilling on or production of oil and gas from land covered by this lease whether or not the well or wells be located on the premises covered by this lease and in such event operations for drilling shall be deemed to have been commenced on said land within the meaning of paragraph 6 of this lease; and the entire acreage constituting such unit or units, as to oil and gas, or either of them, as herein provided, shall be treated for all purposes, except the payment of royalties on production from the pooled unit, as if the same were included in this lease.

(d) For the purpose of computing the royalties to which owners of royalties and payments out of production and each of them shall be entitled on production of oil and gas, or either of them, from the pooled unit, there shall be allocated to the land covered by this lease and included in said unit (or to each separate tract within the unit if this lease covers separate tracts within the unit) a pro rata portion of the oil and gas, or either of them, produced from the pooled unit after deducting that used for operations on the pooled unit. Such allocation shall be on an acreage basis - that is to say, there shall be allocated to the acreage covered by this lease and included in the pooled unit (or to each separate tract within the unit if this lease covers separate tracts within the unit) that pro rata portion of the oil and gas, or either of them, produced from the pooled unit which the number of surface acres covered by this lease (or in each such separate tract) and included in the pooled unit bears to the total number of surface acres included in the pooled unit. Royalties hereunder shall be computed on the portion of such production, whether it be oil and gas, or either of them, so allocated to the land covered by this lease and included in the unit just as though such production were from such land. The production from an oil well will be considered as production from the lease or oil pooled unit which it is producing and not as production from a gas pooled unit; and production from a gas well will be considered as production from the lease or gas pooled unit from which it is producing and not from an oil pooled unit.

(e) The formation of any unit hereunder shall not have the effect of changing the ownership of any delay rental or shut-in production royalty which may become payable under this lease. If this lease now or hereafter covers separate tracts, no pooling or unitization of royalty interest as between any such separate tracts is intended or shall be implied or result merely from the inclusion of such separate tracts within this lease but Lessee shall nevertheless have the right to pool as provided above with consequent allocation of production as above provided. As used in this paragraph 5, the words "separate tract" mean any tract with royalty ownership differing, now or hereafter, either as to parties or amounts, from that as to any other part of the leased premises.

Figure 5.3. — *Continued*

6. (a) If operations for drilling are not commenced on said land or on acreage pooled therewith as above provided on or before one year from this date, the lease shall then terminate as to both parties, unless on or before such anniversary date Lessee shall pay or tender (or shall make a bona fide attempt to pay or tender, as hereinafter stated) to Lessor or to the credit of Lessor in_____ Bank at _____, Texas, (which bank and its successors are Lessor's agent and shall continue as the depository for all rentals payable hereunder regardless of change in ownership of said land or the rentals) the sum of_____ _____ Dollars ($_____), (herein called rentals), which shall cover the privilege of deferring commencement of drilling operations for a period of twelve (12) months. In like manner and upon like payments or tenders annually, the commencement of drilling operations may be further deferred for successive periods of twelve (12) months each during the primary term. The payment or tender of rental under this paragraph and of royalty under paragraph 3 on any gas well from which gas is not being sold or used may be made by the check or draft of Lessee mailed or delivered to the parties entitled thereto or to said bank on or before the date of payment. If such bank (or any successor bank) should fail, liquidate or be succeeded by another bank, or for any reason fail or refuse to accept rental, Lessee shall not be held in default for failure to make such payment or tender of rental until thirty (30) days after Lessor shall deliver to Lessee a proper recordable instrument naming another bank as agent to receive such payments or tenders. If Lessee shall, on or before any anniversary date, make a bona fide attempt to pay or deposit rental to a Lessor entitled thereto according to Lessee's records or to a Lessor, who, prior to such attempted payment or deposit, has given Lessee notice, in accordance with subsequent provisions of this lease of his right to receive rental, and if such payment or deposit shall be ineffective or erroneous in any regard, Lessee shall be unconditionally obligated to pay to such Lessor the rental properly payable for the rental period involved, and this lease shall not terminate but shall be maintained in the same manner as if such erroneous or ineffective rental payment or deposit had been properly made, provided that the erroneous or ineffective rental payment or deposit be corrected within 30 days after receipt by Lessee of written notice from such Lessor of such error accompanied by such instruments as are necessary to enable Lessee to make proper payment. The down cash payment is consideration for this lease according to its terms and shall not be allocated as a mere rental for a period. Lessee may at any time or times execute and deliver to Lessor or to the depository above named or place of record a release or releases of this lease as to all or any part of the above-described premises, or of any mineral or horizon under all or any part thereof, and thereby be relieved of all obligations as to the released land or interest. If this lease is released as to all minerals and horizon under a portion of the land covered by this lease, the rentals and other payments computed in accordance therewith shall thereupon be reduced in the proportion that the number of surface acres within such released portion bears to the total number of surface acres which was covered by the lease immediately prior to such release.

(b) Lessor hereby designates_____ Bank at _____ Texas, and its successors as Lessor's agent to serve as the depository for any payment due with respect to any shut-in gas well. Payment of shut-in gas royalty may be made in the manner provided in paragraph 6(a) hereof for the payment or tender of rentals, including all terms with respect to the deposit of same in the designated depository bank, notwithstanding paragraph 6(a) being otherwise stricken or inoperative due to this lease having a primary term not exceeding one year, if such be the case.

7. If prior to discovery and production of oil, gas or other mineral on said land or on acreage pooled therewith, Lessee should drill a dry hole or holes thereon, or if after discovery and production of oil, gas or other mineral, the production thereof should cease from any cause, this lease shall not terminate if Lessee commences operations for drilling or reworking within ninety (90) days thereafter or if it be within the primary term, commences or resumes the payment or tender of rentals or commences operations for drilling or reworking on or before the rental paying date next ensuing after the expiration of ninety (90) days from date of completion of dry hole or cessation of production. If at any time subsequent to ninety (90) days prior to the beginning of the last year of the primary term and prior to the discovery of oil, gas or other mineral on said land, or on acreage pooled therewith, Lessee should drill a dry hole thereon, no rental payment or operations are necessary in order to keep the lease in force during the remainder of the primary term. If at the expiration of the primary term, oil, gas or other mineral is not being produced on said land, or on acreage pooled therewith, but Lessee is then engaged in drilling or reworking operations thereon or shall have completed a dry hole thereon within ninety (90) days prior to the end of the primary term, the lease shall remain in force so long as operation on said well or for drilling or reworking of any additional well are prosecuted with no cessation of more than ninety (90) consecutive days, and if they result in the production of oil, gas or other mineral, so long thereafter as oil, gas or other mineral is produced from said land or acreage pooled therewith. In the event a well or wells producing oil or gas in paying quantities should be brought in on adjacent land and within three hundred thirty (330) feet of and draining the leased premises, or acreage pooled therewith, Lessee agrees to drill such offset wells as a reasonably prudent operator would drill under the same or similar circumstances.

8. Lessee shall have the right at any time during or after the expiration of this lease to remove all property and fixtures placed by Lessee on said land, including the right to draw and remove all casing. When required by Lessor, Lessee will bury all pipe lines below ordinary plow depth, and no well shall be drilled within two hundred (200) feet of any residence or barn now on said land without Lessor's consent.

9. The rights of either party hereunder may be assigned in whole or in part, and the provisions hereof shall extend to their heirs, successors and assigns; but no change or division in ownership of the land, rentals or royalties, however accomplished, shall operate to enlarge the obligations or diminish the rights of Lessee; and no change or division in such ownership shall be binding on Lessee until thirty (30) days after Lessee shall have been furnished by registered U.S. mail at Lessee's principal place of business with a certified copy of recorded instrument or instruments evidencing same. In the event of assignment hereof in whole or in part, liability for breach of any obligation hereunder shall rest exclusively upon the owner of this lease or of a portion thereof who commits such breach. In the event of the death of any person entitled to rentals, shut-in royalty or royalty hereunder, Lessee may pay or tender such rentals, shut-in royalty or royalty to the credit of the deceased or the estate of the deceased until such time as Lessee is furnished with proper evidence of the appointment and qualification of an executor or administrator of the estate, or if there be none, then until Lessee is furnished with evidence satisfactory to it as to the heirs or devisees of the deceased and that all debts of the estate have been paid. If at any time two or more persons be entitled to participate in the rental payable hereunder, Lessee may pay or tender said rental jointly to such persons or to their joint credit in the depository named herein, or, at Lessee's election, the proportionate part of said rentals to which each participant is entitled may be paid or tendered to him separately or to his separate credit in said depository; and payment or tender to any participant of his portion of the rentals hereunder shall maintain this lease as to such participant. In event of assignment of this lease as to a segregated portion of said land, the rentals payable hereunder shall be apportionable as between the several leasehold owners ratably according to the surface area of each, and default in rental payment by one shall not affect the rights of other leasehold owners hereunder. If six or more parties become entitled to royalty hereunder, Lessee may withhold payment thereof unless and until furnished with a recordable instrument executed by all such parties designating an agent to receive payments for all.

10. (a) The breach by Lessee of any obligation arising hereunder shall not work a forfeiture or termination of this lease nor cause a termination or reversion of the estate created hereby nor be grounds for cancellation hereof in whole or in part. In the event Lessor considers that operations are not at any time being conducted in compliance with this lease, Lessor shall notify Lessee in writing of the facts relied upon as constituting a breach hereof, and Lessee, if in default, shall have sixty days after receipt of such notice in which to commence the compliance with the obligations imposed by virtue of this instrument. The provisions of this paragraph 10(a) shall be applicable to the payment by Lessee of shut-in gas royalty and rentals except that the time for the Lessee to cure any non-payment thereof is otherwise stated in paragraph 4 hereof.

(b) After the discovery of oil, gas or other mineral in paying quantities on said premises, Lessee shall develop the acreage retained hereunder as a reasonably prudent operator, but in discharging this obligation it shall in no event be required to drill more than one well per forty (40) acres, plus an acreage tolerance not to exceed 10% of 40 acres, of the area retained hereunder and capable of producing oil in paying quantities and one well per 320 acres plus an acreage tolerance not to exceed 10% of 320 acres of the area retained hereunder and capable of producing gas or other mineral in paying quantities.

Figure 5.3. — *Continued*

11. Lessor hereby warrants and agrees to defend the title to said land and agrees that Lessee at its option may discharge any tax, mortgage or other lien upon said land, either in whole or in part, and in event Lessee does so, it shall be subrogated to such lien with right to enforce same and apply rentals and royalties accruing hereunder toward satisfying same. Should Lessee become involved in any dispute or litigation arising out of any claim adverse to the title of Lessor to said land, Lessee may recover from Lessor its reasonable and necessary expenses and attorneys fees incurred in such dispute or litigation, with the right to apply royalties accruing hereunder toward satisfying said expenses and attorneys fees. Without impairment of Lessee's rights under the warranty in event of failure of title, it is agreed that if this lease covers a less interest in the oil, gas, sulphur, or other minerals in all or any part of said land than the entire and undivided fee simple estate (whether Lessor's interest is herein specified or not), or no interest therein, then the royalties, delay rental, and other monies accruing from any part as to which this lease covers less than such full interest, shall be paid only in the proportion which the interest therein, if any, covered by this lease, bears to the whole and undivided fee simple estate therein. All royalty interest covered by this lease (whether or not owned by Lessor) shall be paid out of the royalty herein provided. Should any one or more of the parties named above as Lessors fail to execute this lease, it shall nevertheless be binding upon the party or parties executing the same. Failure of Lessee to reduce rental paid hereunder shall not impair the right of Lessee to reduce royalties.

12. When drilling, production or other operations on said land or land pooled with such land, or any part thereof are prevented, delayed or interrupted by lack of water, labor or materials, or by fire, storm, flood, war, rebellion, insurrection, sabotage, riot, strike, difference with workers, or failure of carriers to transport or furnish facilities for transportation, or as a result of some law, order, rule, regulation or necessity of governmental authority, either State or Federal, or as a result of the filing of a suit in which Lessee's title may be affected, or as a result of any cause whatsoever beyond the reasonable control of Lessee, the lease shall nevertheless continue in full force and effect. If any such prevention, delay or interruption should commence during the primary term hereof, the time of such prevention, delay or interruption shall not be counted against Lessee and the running of the primary term shall be suspended during such time; if any such prevention, delay or interruption should commence after the primary term hereof Lessee shall have a period of ninety (90) days after the termination of such period of prevention, delay or interruption within which to commence or resume drilling, production or other operations hereunder, and this lease shall remain in force during such ninety (90) day period and thereafter in accordance with the other provisions of this lease. Lessee shall not be liable for breach of any express or implied covenants of this lease when drilling, production or other operations are so prevented, delayed or interrupted.

13. This lease states the entire contract between the parties, and no representation or promise, verbal or written, on behalf of either party shall be binding unless contained herein; and this lease shall be binding upon each party executing the same, regardless of whether or not executed by all owners of the above described land or by all persons above named as "Lessor", and, notwithstanding the inclusion above of other names as "Lessor", this term as used in this lease shall mean and refer only to such parties as execute this lease and their successors in interest.

IN WITNESS WHEREOF, this instrument is executed on the date first above written.

_____ Lessor
S.S. or Tax I.D. No._____

_____ Lessor
S.S. or Tax I.D. No._____

_____ Lessor
S.S. or Tax I.D. No._____

_____ Lessor
S.S. or Tax I.D. No._____

_____ Lessor
S.S. or Tax I.D. No._____

_____ Lessor
S.S. or Tax I.D. No._____

SINGLE ACKNOWLEDGMENT

THE STATE OF TEXAS,
COUNTY OF

BEFORE ME, the undersigned, a Notary Public in and for said County and State, on this day personally appeared

known to me to be the person whose name   subscribed to the foregoing instrument, and acknowledged to me that   he   executed the same for the purposes and consideration therein expressed.

GIVEN UNDER MY HAND AND SEAL OF OFFICE,
this the   day of   A.D. 19

(L.S.)
My commission expires:

_____
Notary Public in and for the State of Texas

Notary's Printed Name: _____

Figure 5.3.—Continued

SINGLE ACKNOWLEDGMENT

THE STATE OF TEXAS,

COUNTY OF

BEFORE ME, the undersigned, a Notary Public in and for said County and State, on this day personally appeared

known to me to be the person whose name subscribed to the foregoing instrument, and acknowledged to me that he executed the same for the purposes and consideration therein expressed.

GIVEN UNDER MY HAND AND SEAL Of office,

this the day of A.D. 19

(L.S.)

My commission expires:

Notary Public in and for the State of Texas

Notary's Printed Name: ____________

Producers 88 (12 79) Revised
With 320 Acres Pooling Provision

No.____

Oil, Gas and Mineral Lease

FROM

TO

Dated ____________, 19__

No. Acres ________ County, Texas

Term ____

This instrument was filed for record on the ____ day of ________, 19____, at ____ o'clock ____M., and duly recorded in Volume ____, Page ____ of the ________ Records of this office.

County Clerk

By ____________, Deputy

When recorded return to

POUND PRINTING & STATIONERY COMPANY
2325 FANNIN STREET HOUSTON, TEXAS 77002
PHONE 659-3159

CORPORATION ACKNOWLEDGMENT

THE STATE OF TEXAS,

COUNTY OF

BEFORE ME, the undersigned, a Notary Public in and for said County and State, on this day personally appeared

, known to me to be the person and officer whose name is subscribed to the foregoing instrument, and acknowledged to me that the same was the act of the said

a corporation, and that he executed the same as the act of such corporation for the purposes and consideration therein expressed, and in the capacity therein stated.

GIVEN UNDER MY HAND AND SEAL OF OFFICE, this the day of A.D. 19

(L.S.)

My commission expires:

Notary Public in and for the State of Texas

Notary's Printed Name: ____________

Figure 5.3.—*Continued*

Another function of the landman is the possible negotiation with other operators for the pooling of production in a field. An operating company does not usually want to spend the money to drill an exploratory well without having the right to drill additional wells on adjacent property in the event that the wildcat is successful. Depending on well spacing regulations in the area, the company will lease enough land from individual owners or unitize enough drilling units with other producers to develop a significant part of the field if the prospect is promising. The terms *pooling* and *unitizing* both refer to the activity of combining more than one lease for the purpose of more efficiently producing a reservoir (fig. 5.4).

Many other types of joint agreements are used to finance proposed petroleum investments. *Support agreements* take place when a company or individuals provide support moneys for the privilege of obtaining information from another company's exploratory well. Testing and drilling of that well can provide valuable data about the formation that the other company may need if it is investigating the same or a similar formation. In the case of *dry hole money*, the contributor agrees to pay so much per foot drilled (not to exceed a stipulated sum) if the well does not produce. If the contributor agrees to pay even if the well does produce for the other company, then the funds are called *bottomhole money.*

Dry hole and bottomhole contributions are pledged early in a prospect for exploration. Other types of joint arrangements call for complex accounting procedures later on, especially if the well produces and a field is developed. An operating agreement is drawn up that usually names one of the working-interest owners as operator and details how expenses and revenues, if they occur, are to be apportioned.

In order to pay for exploration and drilling, the operator will estimate expenses and prepare an AFE (authority for expenditure) to send for signature to each nonoperator with a working interest. (The operator will probably continue to send out AFEs for subsequent activities on the well.) If production occurs, the dividing of royalties and working interests entails the use of various formulas. Besides royalty payments off the top, *overriding royalties* may exist. An overriding royalty is a percentage of the lessee's working interest that has been awarded to a party. Like the landowner's royalty, the overriding royalty is free from all production costs. A *division order* lists all interest owners in a given unit and stipulates their individual net revenue interests as decimal figures of the total, thus allowing for payments.

Although the basic principles of offshore leasing are similar to those of onshore leasing, one major difference exists. Operating companies lease, in the case of the outer continental shelf, individual blocks from the federal government rather than from private landowners. Usually, private industry has performed geological and seismic exploration in an area and recommended to the government the leasing of promising tracts. The Minerals Management Service (MMS) of the U.S. Department of the Interior manages the leasing process. Companies bid competitively for the right to spend millions of dollars to explore an area

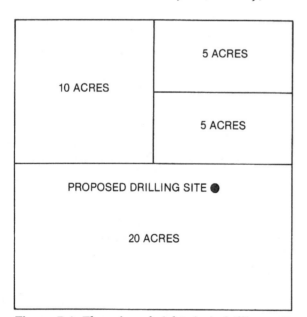

Figure 5.4. The mineral rights in a drilling unit consisting of forty acres may be owned by different individuals. The surface rights may be owned by yet other individuals.

deemed to have potential. The royalty amount paid on federally owned lands, typically a sixth, has traditionally been a higher percentage than that paid to individuals.

Obtaining the right to drill is handled very differently in other countries where mineral rights, and the petroleum industry itself, may be owned by the government of a country. In many cases, marketing of crude oil, products, and natural gas actually takes place between governments. In such situations, the economic activity can be radically different from that in the United States.

EVALUATION OF ESTIMATED RESERVES

In order to weigh an investment, a prediction or estimate of producible reserves is usually made before drilling. Geophysical surveys and geological data enable the geologist to estimate the area of the reservoir and the possible thickness of its pay zone in feet. By multiplying the two figures, they are able to predict the number of acre-feet in the reservoir, and, assuming a porosity, its pore-space volume. Until various production tests are done after drilling, however, predicting the amount of production possible at a particular drill site is difficult. Also, the drilling itself will further define the extent of the field and determine the number of commercially productive zones that can be penetrated in reaching total depth. Geologists or production engineers will also take into account state regulations determining well spacing for the area. Thus they are able to estimate the time it would take for the maximum recovery of reserves.

At this time the economic analysts must convert the estimated reserves figure into financial terms. They give a price per barrel to the predicted production. The price will depend on all the factors: accessibility to market, world market pricing, potential political situations, fluctuations in world supply, and fluctuations in consumption. But before the estimated revenues, given

now in dollars, are balanced against the costs of drilling in a cash flow chart, the analysts must take two other factors into account: regulation and taxation.

REGULATION CONSIDERATIONS

After the discovery of the East Texas Field in 1930, an oversupply of oil occurred. In order to prevent marketing chaos and waste of those hydrocarbons still in the ground, regulation seemed to be needed. Various state regulatory agencies were created. Most of them set some sort of proration policies. *Proration* at that time involved regulation of the amount of statewide production according to the market needs of purchasers of oil and gas. The *allowable*, the amount allowed to be produced, was prorated among the well owners in each field. The regulatory body limited production amounts unless capacity production was needed in the market. Today proration policies set allowables based on the ability of individual fields, not on market needs.

To prevent wasting of their natural resources, various states have regulatory bodies, such as the Office of Conservation in Louisiana, the Oil Conservation Division in New Mexico, the Division of Oil and Gas in California, and corporation commissions in Kansas and Oklahoma. Because Texas is important as an oil-producing state, the functions of its oil and gas regulatory body—the Railroad Commission—are used as an example.

The chief responsibilities of the Oil and Gas Division of the Railroad Commission of Texas are as follows:

1. To regulate the quantity of oil or gas flowing from a tract (fig. 5.5)

2. To determine well spacing (fig. 5.6)

3. To authorize or force pooling of acreage (interests) for conservation purposes

4. To exercise pollution control over production and transportation activities by establishing adequate safeguards

RAILROAD COMMISSION OF TEXAS
OIL AND GAS DIVISION
Gas Well Allowable Supplement

The allowable set forth below supersedes all previous allowables set for this well irrespective of the manner in which such previous allowables were set, unless otherwise noted under "Remarks."

This instrument constitutes authority to produce the well described in compliance with the Statutes of Texas, and under the Rules, Regulations and Orders of the Commission applicable thereto from the effective date shown below until the issuance of a new schedule in which the allowable for this well is included.

All gas produced from gas wells must be measured by recording meters. All Gas Volumes are in Thousands of Cubic Feet (MCF) at 14.65 Lbs. per Sq. In. Pressure, and Flowing Temperature of 60° Fahrenheit.

DIST. NO.	RRC IDENT. NO.	COUNTY NAME		SUPPLEMENT NO.	DATE OF ISSUANCE	CURRENT STATUS	STATUS DATE	OPERATOR NO.

FIELD NAME			FIELD NUMBER	LEASE NAME			WELL NO.

MONTH DAY YEAR	PREVIOUS ALLOWABLE	NEW ALLOWABLE	†	POTENTIAL	Deliverability	BHP OR SIWHP	ACRES	ACRE−FEET	PRODUCTION

REMARKS:

* INDICATES CHANGE AND REASON FOR SUPPLEMENT

† SEE REASON BELOW
   1 – INITIAL ALLOWABLE ASSIGNMENT
   2 – REVISED ALLOWABLE DUE TO OVERPRODUCED LIMITED WELL
   3 – CHANGE OF @ LIMITATION
   4 – CHANGE OF @ LIMITATION – BUT RESTRICTED TO TEST
   5 – SUBMIT AVERAGE PRODUCTION ON FORM G–2

By order of the Railroad Commission of Texas

SEND INQUIRIES TO:
DIRECTOR, PRODUCTION and PRORATION
Oil and Gas Division

Figure 5.5. When the proper gas well completion and test forms have been filed and accepted, the Gas Well Allowable Supplement form is used by the commission to advise the operator, gas gatherer, and district office of the initial gas allowable. (Courtesy of Railroad Commission of Texas)

5. To set conservation standards for and to regulate additional recovery projects
6. To formulate proration policies[2]

In order to prevent waste during production, the Railroad Commission has established over the years a series of rules concerning well spacing, density, and allowables. An allowable is based on depth of an individual well and size of the unit assigned. The allowed production is often referred to as the *maximum efficiency rate*

(MER), since the maximum yield of a field possible without damage or waste to the field is the objective behind the yardsticks created over the years to define allowables. Draining a field too fast or spacing wells too closely can diminish the ultimate production of a field.

The laws on well spacing are designed to help an operator develop a field in an efficient and timely manner that will be as economically desirable as possible without damaging the reservoir. State regulatory agencies, as well as operators, are aware of

---

2. Megill, 14.

the relationship between the cost of drilling the wells in a field, the volume and value of their ultimate production, and the resulting productive life period. This relationship, as well as state law, influences how many wells should be drilled when a field is developed. How many wells and where they should be placed become critical for both the economic health of the operator and the future productive health of the reservoir.

The state is also concerned with well spacing in terms of protecting its individuals' property rights. Since the same reservoir can be drained by two or more competitive operators, the well spacing rules try to handle this situation as fairly as possible without interrupting the inevitable competition unduly.

While the states handle prorationing and well spacing, the federal government, within the last two decades, has become increasingly influential in the regulation of oil and gas. The industry has operated under a variety of governmental regulations, tax incentives, and research and development assistance plans. Since the enactment of President Jimmy Carter's National Energy Plan and the Natural Gas Policy Act (NGPA), the prevailing emphasis of the United States has been one of deregulation. Price controls of U.S. oil were removed in 1981. Wellhead prices for oil have been allowed to rise to world levels to encourage

▶ Return each W-1 with plat and $100.00 fee.
Make a check or money order payable to the State Treasurer of Texas. Address to:
Railroad Commission of Texas
Oil and Gas Division, Drilling Permits
P. O. Drawer 12967, Capitol Station
Austin, Texas 78711
▶ File a copy of W-1 and plat in RRC District Office.

RAILROAD COMMISSION OF TEXAS
Oil and Gas Division
Application for Permit to Drill, Deepen, Plug Back, or Re-Enter

▶ Read Instructions on Back

Form W-1
Rev. 9/1/83
483-060

Purpose of filing (mark appropriate boxes): [X] Drill [ ] Deepen (below casing) [ ] Deepen (within casing) [ ] Plug Back [ ] Re-Enter
[ ] Directional Well [ ] Sidetrack [ ] Amended Permit (enter permit no. at right & explain fully in Remarks)

Enter here, if assigned: ▶ API No. 42- ; Permit No. ▶ ; Rule 37 Case No. ▶

1. Operator's Name (exactly as shown on Form P-5, Organization Report): BREAKER OIL COMPANY
3. RRC Operator No.: 235041
4. RRC District No.: 09
5. County of Well Site: WISE
2. Address (including city and zip code): 302 MAIN, DALLAS, TX 79701
6. Lease Name (32 spaces maximum): F. T. ROBERTS UNIT
7. RRC Lease ID No.:
8. Well No.: 2
9. Total Depth: 6000'
10. Location
• Section ______ Block ______ Survey JAMES THORNTON Abstract No. A- 1852
• This well is to be located 4 miles in a NW direction from DECATUR which is the nearest town in the county of the well site.

11. Distance from proposed location to nearest lease or unit line 700 ft
12. Number of contiguous acres in lease, pooled unit, or unitized tract 117 (OUTLINE ON PLAT.)

13. FIELD NAME (Exactly as shown on RRC proration schedule). List all established and wildcat zones of anticipated completion. Attach additional Form W-1's as needed to list these zones. One zone per line.	14. Completion depth	15. Spacing pattern (ft.)	16 Density pattern (acres)	17 Number of acres in drilling unit for this well OUTLINE ON PLAT.	18 Is this acreage assigned to another well on this lease & in this reservoir? If so, explain in Remarks.	19. Distance from proposed location to nearest applied for, permitted, or completed well, this lease & reservoir (ft.)	20 Oil, gas, or other type well (Specify)	21 No. of applied for, permitted, or completed locations (including this one) on lease in this reservoir. OIL	GAS
WILDCAT ABOVE 6000'	6000'	467-1,200	40	40	NO	N/A	GAS	--	--
BOONSVILLE (BEND CONGL. GAS)	4355'	467-2,000	320/160	117	NO	N/A	GAS	--	--
BOONSVILLE (CONS. CONGL.)	5800'	660-1,320	80	80	YES	1,100	GAS		1

22. Perpendicular surface location from two nearest designated lines:
• Lease/Unit 700' FROM EL & 1,250' FROM NL
• Survey/Section 700' FROM EL & 1,400' FROM SL

If a directional well, show also projected bottom hole location:
• Lease Unit ______
• Survey Section ______

23. Is this a pooled unit? Yes [X] ▶ (Attach Form P-12 and certified plat.) No [ ]
24. Is Item 17 less than Item 16 (substandard acreage for any field applied for)? Yes [X] ▶ (Attach Form W-1A) No [ ]
25. Is this wellbore subject to Statewide Rule 36 (hydrogen sulfide area)? Yes [X] No [ ]
If subject to Rule 36, is Form H-9 filed? Yes [X] No [ ] ▶ If not filed, explain in Remarks.
26. Do you have the right to develop the minerals under any right-of-way that crosses, or is contiguous to, this tract? Yes [X] No [ ]
If not, and if the well requires a Rule 37 or 38 exception, see Instructions for Rule 37.

Remarks: WELL #1 IS IN BOONSVILLE (CONSOLIDATED CONGL.) BUT WILL BE PLUGGED IF THIS IS A GOOD COMPLETION.

I certify that information stated in this application is true and complete, to the best of my knowledge.
Signature: Joe Terry — Name and title of operator's representative: JOE TERRY, DISTRICT ENGR.
Date: 09 (mo.) 01 (day) 83 (yr) — Tel.: 214 (Area Code) 555-1212 (Number)

• RRC Use Only •

Figure 5.6. Form W-1 shows whether or not the operator is complying with the commission's regulation for well spacing in a particular field. (Courtesy of Railroad Commission of Texas)

further exploration and development, but increased revenues are returned to the federal government in the form of the windfall profit tax. And, an elaborate set of rules is slowly allowing the phasing out of many natural gas price controls at the wellhead.

In spite of the trend toward deregulation, a tremendous amount of federal regulation continues to affect the oil and gas industry. A new awareness by the American public of environmental concerns and the limited supply of hydrocarbons has renewed interest in regulation. The Department of Energy (DOE) was formed in 1977 to deal with energy policy, resource development, land usage, and other issues. The Federal Power Commission, which regulated interstate gas pipelines and pricing, was reorganized as the Federal Energy Regulatory Commission (FERC). The Department of the Interior (DOI) decides the timing and amounts of federal lease sales.

TAXATION CONSIDERATIONS

Federal influence is felt perhaps most strongly in taxing provisions for the oil and gas industry. The three basic kinds of taxes that affect the petroleum industry are production, ad valorem, and income taxes. *Production taxes* are state and municipal taxes on oil and gas products, levied at the wellhead for the removal of the hydrocarbons. They are sometimes referred to as *severance taxes. Ad valorem taxes* are state or county taxes based on the value of a property. It is, however, the federal *income tax* that most strongly influences the setting up of a cash flow analysis to aid geologists and managers in making the decision of whether or not to drill.

Many income tax provisions take the form of incentives to encourage the petroleum industry to take the enormously expensive risks necessary in exploration. Very important among these incentives are the concepts of depletion and depreciation. Since these two taxing advantages greatly influence managerial decisions, a geologist needs to understand the ways in which they are used in the timing of money programs.

Also important among taxation considerations are the choices a company makes concerning whether to expense or to capitalize various drilling and development costs. Drilling expenses can be categorized in several ways under the Internal Revenue Code. The choices that a company makes in this area affect the amount of capital it will have either early or late in the development of a prospect. The basic concept behind this sort of timing of the availability of capital is the separation of drilling costs into capitalized and expensed categories. *Expensed* costs are deducted from income in the year in which the expenditures are incurred. *Capitalized* costs are deducted from income over the years of useful life of an item purchased. Items most often capitalized and depreciated are those that have potential salvage value.

Although equipment is tangible, about 70% of drilling and development costs are considered intangible. *Intangible development costs* (IDC) are costs of items that do not have a salvage value, such as site preparation, rig transportation, rig operation, drilling fluid, formation tests, cement, well supplies, and other expenses relating to activities on the rig. Since these costs represent items that will not be recoverable for future projects, they have no depreciation value from a tax standpoint.

The reasoning behind the designations is that early deduction is best for a company when it is possible by law. Therefore, a company usually expenses all items possible in the year in which they are incurred. These items usually include intangible expenses. Items that cannot be expensed are capitalized. Capitalized expenditures, or noncash items, can be deducted over the years through *depreciation* and *depletion*.

Depreciation is an annual reduction of income reflecting the loss in useful value of capitalized investments by reason of wear and tear.... Depletion is a reduction in income reflecting the exhaustion of a mineral deposit. It is allowed to prevent the taxation of a capital asset as ordinary income.[3]

---

3. Megill, 33.

TABLE 5.1

Typical Tax Treatment Allowed By IRS For Project Costs

Expensed	Capitalized	Expensed or Capitalized
Dry holes	Tangible costs (equip-	Core drilling
Intangible costs	ment, etc.) for success-	Seismic crews
Unoperated leases	ful wells	Data processing
(surrendered)	Lease bonus	Interpretation
Rentals and taxes		Test well contributions
		Portion of overhead

These two taxing concepts do not provide for a return on money as speedy as expensing items, but they do provide a means of deduction over the years of a project. Tangible drilling costs, such as casing and rig equipment, must by law be capitalized. However, they can still be depreciated.

Two different mathematical methods used by tax accountants to figure depletion are referred to as the *cost* and the *percentage* methods. The method chosen by the accountants is the one providing the highest deduction, if percentage depletion is allowed in the particular case.

Table 5.1 shows how different costs are designated for income tax purposes. After such designations are made, the equipment is depreciated, and percentage or cost depletion is subtracted from remaining taxable income if possible. At present, the depletion allowance is generally reserved for companies with only small amounts of oil production.

CASH FLOW ANALYSIS

After the economic analysts decide how to designate drilling and development expenses for tax purposes, a cash flow chart can be prepared. This comparison of predicted expenses and predicted income will often analyze an entire field's future development. The difference between inflows and outflows of funds over a period of years is called *net cash flow*. All figures, of course, represent estimates. By predicting a cash flow stream for a number of years, geologists and managers can evaluate individual investments and compare several investment opportunities at once.

In order to prepare the side of the ledger predicting outgoing funds, the analysts must try to determine how much men, material, and operations are going to cost over a period of years. First, the number of exploratory and development wells must be predicted. Each must have a drilling rig, attendant equipment and crew, and, for land wells, the associated grading of the location, pits, and access road. Casing and cementing services may be needed. Drilling mud and oil for the mud is a big expense. Water and fuel to run the rig will have to be paid for. Various tests such as logging and drill stem tests will have to be paid for. Third-party service companies will probably be called in to complete the well or plug it, with possible accompanying costs for cement and casing.

Predicting costs is more difficult for offshore drilling than for onshore drilling because of the large investments involved and the much greater sums dealt with in case of hole problem. Costs increase with water depth and weather severity. A crucial factor is whether mobile drilling rigs and seafloor producing units will be used or a fixed platform will be built. The distance from shore affects expenditures. If climatic conditions are severe, as in the offshore Alaskan, Canadian, and North Sea areas, the cost for a production system may be three times as much as for a similar one in a moderate-cost area such as the Gulf of Mexico. Extreme cold, shortness of equipping time, and the danger of floating icebergs and breaking-up ice require innovative technology with enormous costs. Often, estimates of the price of such technology take the form of educated guesses.

TABLE 5.2

"INVESTMENT" CASH FLOW STREAM (AFTER FEDERAL INCOME TAX) – $M
SCHOOL PROSPECT

Year	Outflows			Inflow	Net outflow
	Capitalized	Tangible	Total intangible and expenses	Income tax credit	Investment cash flow (after FIT)
0	-500	–	–	–	-500
1	-100	–	-200	+100	-200
2	–	–	-1200	+600	-600
3	–	-200	-800	+400	-600
4	–	-500	-1800	+900	-1400
5	–	-100	-800	+400	-500
Total	-600	-800	-4800	+2400	-3800

SOURCE: Robert Megill, *An Introduction to Exploration Economics,* 2d ed., Copyright PennWell Books, 1979, p. 56.

Once the predicted expenses have been estimated as accurately as possible, a number of tables can be constructed to document the final subtraction of expenses from gross inflow (tables 5.2–5.5).

In the method illustrated in table 5.2, the investment cash flow stream, or expenditures of the first 5 years of a project, is developed first. The expenses of the project are shown by their various designations for income tax purposes. The early outflows reflect money spent for leasing, geophysical work, dry holes, a discovery well, development wells, and lease facilities. The inflow column reflects credit for income tax saved by the expensing of all items possible.

Table 5.3 represents the income cash flow stream from years 3 through 13 of the project. The income reflects mostly the sale of oil, gas, and natural gas liquids, and tax credits from expensed items, depletion, and depreciation. The outflows reflect the royalty, production and ad valorem taxes, operating expenses, employee benefits, and overhead. Column 7 shows the income cash flow before federal income taxes are taken out.

Table 5.4 shows the subtraction of federal income tax from the income cash flow arrived at in table 5.3. From column 7 in table 5.3 are subtracted the money from surrendered leases, depreciation, and depletion to arrive at the company's taxable income on this particular project. The tax itself is stated in column 11. Column 11 is subtracted from column 7 to show in column 12 the actual income cash flow for each year. From this final figure will be subtracted the outflow figured in table 5.2. That calculation is illustrated in table 5.5.

Table 5.5 shows the resulting net cash flow after federal income tax is subtracted. It is this final figure for the project that a company uses with other evaluation tools in comparing this particular investment to other similar ones. The last column, the cumulative net cash flow, simply brings the reader up to date about the total amount amassed so far at the end of each year.

Future gross earnings for a project can thus be predicted. The cost of a project can also be predicted. Then if the cost is subtracted from gross earnings, net earnings can be predicted. But another crucial unknown exists. For a company to compare investment alternatives, it must figure a rate of return for each of its contemplated ventures. If the rate of return represents simply the ratio between an initial investment and the investment's predicted earnings, the

TABLE 5.3

"INCOME" CASH FLOW STREAM – M$
SCHOOL PROSPECT

	1	2	3	4	5	6	7
Year	Gross revenue	Royalty	Net revenue (1–2)	Prod. & Ad Val. tax	Dir. oper. exp.	Ovhd. above Ise.*	Inc. cash flow before FIT (3–4–5–6)
0–3							
4	2,890	480	2,410	240	60	20	2,090
5	2,890	480	2,410	240	60	20	2,090
6	3,160	530	2,630	260	60	20	2,290
7	3,160	530	2,630	260	60	20	2,290
8	3,160	530	2,630	260	60	20	2,290
9	1,560	260	1,300	130	30	10	1,130
10	1,560	260	1,300	130	30	10	1,130
11	1,560	260	1,300	130	30	10	1,130
12	1,560	260	1,300	130	30	10	1,130
13	1,560	260	1,300	130	30	10	1,130
Total	23,060	3,850	19,210	1,910	450	150	16,700

*Includes employee benefit expense
SOURCE: Robert Megill, *An Introduction to Exploration Economics,* 2d ed., Copyright PennWell Books, 1979, p.60.

TABLE 5.4

"INCOME" CASH FLOW STREAM (AFTER FEDERAL INCOME TAX) – M$
SCHOOL PROSPECT

	7	8	9	10	11	12
Year	Income cash flow BFIT	Surr. leases	Deprn. & depln.	Taxable income (7–8–9)	Federal income tax	Income cash flow AFIT (7–11)
0–3						
4	2,090	200	110	1,780	890	1,200
5	2,090	200	110	1,780	890	1,200
6	2,290		110	2,180	1,090	1,200
7	2,290		110	2,180	1,090	1,200
8	2,290		110	2,180	1,090	1,200
9	1,130		70	1,060	530	600
10	1,130		70	1,060	530	600
11	1,130		70	1,060	530	600
12	1,130		70	1,060	530	600
13	1,130		70	1,060	530	600
Total	16,700	400	900	15,400	7,700	9,000

SOURCE: Robert Megill, *An Introduction to Exploration Economics,* 2d ed., Copyright PennWell Books, 1979, p. 61.

TABLE 5.5

NET CASH FLOW (AFTER FEDERAL INCOME TAX) – $M
SCHOOL PPROSPECT

Year	"Investment" cash flow	"Income" cash flow	Net cash flow (after FIT)	Cumulative net cash flow	
0	−500		−500	−500	
1	−200		−200	−700	
2	−600		−600	−1,300	
3	−600		−600	−1,900	
4	−1,400	1,200	−200	−2,100	Maximum negative
5	−500	1,200	700	−1,400	cash flow, − $2,100
6		1,200	1,200	−200	
7		1,200	1,200	1,000	Cumulative net cash
8		1,200	1,200	2,200	flow becomes posi-
9		600	600	2,800	tive during year 7.
10		600	600	3,400	
11		600	600	4,000	
12		600	600	4,600	
13		600	600	5,200	
Total	−3,800	9,000	5,200		
					Actual value profit: + $5,200

SOURCE: Robert Megill, *An Introduction to Exploration Economics,* 2d ed., Copyright PennWell Books, 1979, p. 62.

calculation is not taking into account the time value of money. Therefore, more complicated methods and their formulas are used to figure a rate of return that takes into consideration the value of a sum of money today and the value of that same amount, *if invested,* in the future.

PRESENT VALUE CONCEPT

When a cash flow analysis is performed to help geologists and managers in their investment decisions, one of the most important concepts operating is that of the *present value* of money. Crucial to an understanding of investment economics is the idea that an amount of money predicted for gain in a certain number of years is worth less than that amount today. When the present value of a predicted amount of earnings is calculated, the amount is said to have been *discounted.*

One way of understanding discounting is to realize that it is the reverse of *compound-*

ing. We are familiar with the process of accruing interest on saved or invested moneys. One dollar invested at 15% compounded annually would be worth $4.05 in 10 years. In other words, $4.05 received 10 years from now is worth only $1.00 today if we have a 15% compounded annual interest. The present worth of a dollar at some future date is the reciprocal of the future value of a dollar invested today for the same time, at the same rate, and with the same compounding interval. Table 5.6 shows how $1.00 is compounded to $4.05 and how $4.05 is discounted to $1.00.

When financial planners predict future net earnings, they figure the earnings' present value. The question is which percentage rate to use in order to discount the future funds. The chosen discount rate has a definite relationship to the potential rate of return possible for a project's investment. The investment sum's rate of return represents the relationship between the amount invested in year zero and the amount earned in future years. The discount rate that makes the present value of future net earnings the same as the amount of the initial investment is called the *discounted cash flow rate of return* (DCFR). This rate becomes the quoted rate of the project.

How is a rate of return figured after predicted reserves and their market prices are estimated? The calculation is a trial-and-error method. A discount rate is chosen at random and used to discount all predicted cash flows back to time zero – the present year, when the investment is going to be made. Thus, the present value of all the future cash revenues generated is established according to the chosen rate. If the resulting present value figure exceeds the investment, the discount rate selected was too low. If the present value figure is lower than the initial investment, the discount rate selected was too high, and a smaller discount rate should be tried. When a discount rate that makes the present value of the estimated earnings equal to the investment under consideration is found, then that discount rate becomes the rate of return predicted for that particular investment.

In working with the huge sums required in oil exploration and also with the diverse number of years required for the development of a field, the present value concept becomes very important. For example, assume that a similar amount of money can be made from production in the second and

the fourth years of the life of a well. Planners might have to choose which year to bring in the production. They will know that the sum of money made in the second year will be worth more according to the present value concept, all other considerations being equal. Also, using the present value concept, companies can compare rates of return of different ventures and know that this method takes into consideration the time implications of money.

Rate of return is only one of many yardsticks used for defining the economic worth of a prospect. *Yardstick* is the term used for any of the many evaluative tools and processes managers and geologists employ in their decision analysis to compare various projects. Many of the yardsticks are concerned with that fascinating area of exploration – risk and probability.

RISK ANALYSIS

Risk analysis may be defined in the oil business as the activity of assigning probabilities to all the possible outcomes of a drilling venture. So many critical variables are involved that intuition no longer will suffice in the face of the huge expense that escalating costs and new technology have brought to the oil industry today. Fortunately, computers through modeling and simulation have made possible the stringing

TABLE 5.6

COMPOUNDING AND PRESENT VALUE CALCULATIONS

Compound Interest Calculation			Present Value Calculation		
Year	Beginning Amount	Ending Amount	Year	Beginning Amount	Ending Amount
1	$1.00	$1.15	10	$4.05	$3.52
2	$1.15	$1.32	9	$3.52	$3.06
3	$1.32	$1.52	8	$3.06	$2.66
4	$1.52	$1.75	7	$2.66	$2.31
5	$1.75	$2.01	6	$2.31	$2.01
6	$2.01	$2.31	5	$2.01	$1.75
7	$2.31	$2.66	4	$1.75	$1.52
8	$2.66	$3.06	3	$1.52	$1.32
9	$3.06	$3.52	2	$1.32	$1.15
10	$3.52	$4.05	1	$1.15	$1.00

TABLE 5.7

EXPECTED VALUE CONCEPT

Outcome	Worth		Probability		Expected Value
Dry hole	$-1 MM	×	.900	=	$-0.9 MM
Marginal field	$ 2 MM	×	.025	=	$ 0.05 MM
Successful field	$ 40 MM	×	.075	=	$ 3 MM

together of many variables and their probability distributions.

It is possible on a simple scale to calculate some of the chance that exists in all drilling ventures. Central to the use of probabilities are certain statistics stating how frequently a dry hole, a marginal well, or a giant discovery occurs in any given formation. The statistics used by any one company may have been derived from the company's own past performances or may have been taken from various tables reflecting more wide-ranging data. Assume that fairly accurate statistics can be obtained and focus on how they are manipulated to help in decision making.

A process using the *expected value concept* is one that multiplies the expected gain of a possible outcome, such as drilling a successful discovery well, by the probability of its occurrence. This product is called the expected value. For example, assume that a cash flow chart has shown the potential net earnings of a 20-million-barrel discovery at a certain price in the future to be $40 million. Statistics indicate a 7.5% chance of making such a discovery. Therefore, the expected value would be figured in the following manner:

.075 × $40 MM (million) = $3 MM

A contemplated exploratory well, of course, has another possible outcome– a dry hole. Also, degrees of success vary. The new field could be capable of producing 1 million barrels of oil or 20 million. How does the expected value concept give an average predictability on a sum that might be earned from a certain decision? Usually, a geologist will predict at least three outcomes concerning the size of a reservoir–a most likely case and the high and low

extremes. The further usefulness of the expected value theory can be shown by working with the three outcomes just mentioned (table 5.7).

The expected value of the dry hole is $-0.9 MM. The expected value of the marginal well is $0.05 MM. The expected value of the successful well is $3 MM. (The dry hole is a negative number because of the funds that went into drilling it.) These figures can be used to find an expected value for the project as a whole by finding the algebraic sum of the expected values of all possible outcomes (fig. 5.7).

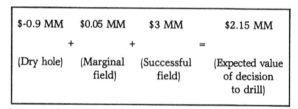

Figure 5.7. Algebraic sum of expected values

If a company is using the expected value concept to compare alternative decisions, the company will choose the action with the highest expected value. This concept can also be used to compare various projects with each other. However, the concept will not work unless the company uses it to measure similar alternatives and uses it consistently. The expected value concept is a strategy for repeated use rather than an absolute measure of the profitability of any one venture. If a decision maker consistently uses this philosophy and consistently chooses the alternative having the highest expected value, the total net gain from all decisions will be higher than if he does not. Most sophisticated forms of risk analysis are based on this principle. The weighted

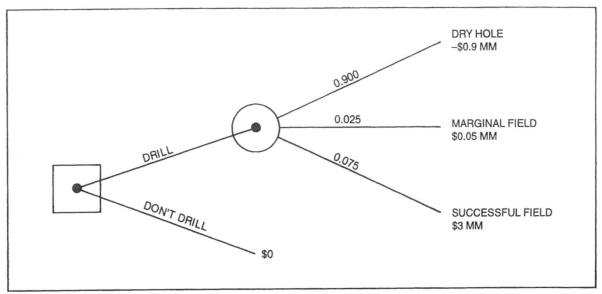

Figure 5.8. Decision tree for an example project. Calculated expected value for decision to drill is $2.15 MM.

choices in the example are shown in the form of a decision tree in figure 5.8. A more detailed decision tree of another project is shown in figure 5.9.

The decision makers go through the processes of estimating reserves, evaluating them, predicting costs, finding net earnings, establishing present value for the earnings and thus a rate of return on the investment under consideration, and finally weighing the risks. Geologists need to be familiar with these processes, since decisions based on geological data are so influenced by the fluctuations of financial considerations.

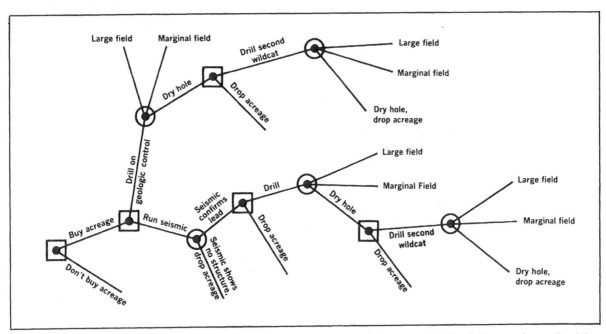

Figure 5.9. More involved decision tree for another project (Courtesy of Paul Newendorp, *Decision Analysis for Petroleum Exploration,* Copyright PennWell Books, 1975, p. 117.)

6

WELL SITTING

The commitment to drill an exploratory well sets in motion a long chain of events. Well plans are engineered. The site is surveyed and prepared for the drilling rig (fig. 6.1). Contracts are secured for drilling, casing, cementing, coring, logging, and testing services. Arrangements for supplies of all types are made. And a geologist is assigned to the well. The geologist who in-itially brought the location to the attention of others may continue to work at the well site, or someone new may be assigned. The geologist may be a partner in the venture or one of a staff of hundreds of geoscientists in a giant corporation. Whatever the operator arrangements, the geologist will have clearly defined responsibilities for certain tasks, reports, and decisions at the well site.

Figure 6.1. Drilling a Wildcat

TABLE 6.1

INFORMATION CONCERNING AN EXPLORATORY WELL

General information	
Location	South central Scurry County, Texas
Regional well density	One well/300 square miles
Nearest wells	Well 6 miles NW drilled to San Andres formation
	Well 14 miles SW drilled to Atoka shale
Reason for location	Subsurface trend
	Seismic closure of 250 feet over 2,000 acres
Acreage leased by company	4,000 acres in area, including almost ¾ of the 2,000-acre structure being tested
Geologic column anticipated	
0'–1600' (1600')	Non-marine red beds, salt, anhydrite
1600'–2600' (1000')	San Andres limestone
2600'–3650' (1050')	Clear Fork limestone
3650'–5200' (1550')	Wolfcamp limestone
5200'–7000' (1800')	Wolfcamp lime and shale
7000'–7800' (800')	Pennsylvanian limestone
7800'–7900' (100')	Atoka shale
7900'–8050' (150')	Mississippian lime and chert
8000'–8450' (450')	Ellenburger limestone
8450'	Granite basement
Objectives	
Primary	Testing Ellenburger
Secondary	Testing all formations below 1600' if indicated, except Atoka shale and Mississippi lime
Drilling and casing program	Drill 12¼" hole to San Andres using saltwater mud.
	Cement 9⅝" surface casing.
	Drill 8¾" hole to total depth (TD) using low water loss mud; possible circulation losses in lower Wolfcamp.
	Cement 7½" oil string to complete.
Estimated time	
Drilling	42 days
Testing	11 days
Evaluation program	
Cuttings samples	20' intervals, 0'–1600'
	10' intervals, 1600'–TD
Mud logging	9⅝" casing shoe to TD
Wireline logging	
Caliper and dipmeter	9⅝" casing shoe to TD
SP, dual induction, laterolog, microlog	9⅝" casing shoe to TD
Gamma ray and neutron	0'–TD if oil string is cemented in
Coring	Top 60' of Pennsylvanian limestone
	Top 100' of Ellenburger
Drill stem test	All lost circulation zones
	All shows

DUTIES OF THE GEOLOGIST

The geologist prepares for the assignment by gathering materials and supplies, alerting technical support services, and reviewing company procedures. Assume, for example, a hypothetical prospect in West Texas. The operating company might provide the geologist with the information shown in table 6.1.

Before leaving for the rig, the geologist will study samples of the rocks that will be drilled through. The samples may be from outcrops or from other wells. On site, the geologist must quickly become familiar with the crew, the operation, particular quirks of each piece of equipment, and anything else that may affect the quality of the data to be gathered. One essential item is to confirm the elevation at the rig floor. An error in the elevation will affect all the logged data.

The geologist usually supervises the mud logger, who collects cuttings at the shale shaker (fig. 6.2). Normally the mud logger will log lithology after examining the cuttings. He also keeps watch on the circulation system and drilling fluid properties for the mud engineer. The geologist may also collect cuttings and prepare a lithology log, but often just reviews the mud log.

Besides filling out daily and weekly reports, the geologist must be prepared to assist the engineers at the rig with downhole pressure evaluation, coring procedures, wireline logging operations, and determination of depths for setting casing strings. The geologist may also be called upon to help with velocity-shot surveys for the geophysicists. And the geologist usually recommends whether to plug and abandon the well or to complete and prepare for production.

Figure 6.2. Shale shakers

Figure 6.3. Cuttings

LOGGING

Sample Logs

The modern geologist seldom spends full time at the drill site unless the well is a remote or an expensive offshore one. But the duty to log the lithology of a well is fundamental, so the geologist must supervise the operation. Lithological evidence comes from samples—cuttings, sidewall samples, and cores—and from wireline logs.

Cuttings. Samples of cuttings (fig. 6.3) from the well are taken from the shale shaker at definite intervals, say every 10 feet. The geologist must keep in mind that even though the drill bit has moved down 10 feet, and the time it takes the mud to circulate to the surface is known, the cuttings may not be carried at the same rate as the fluid. A certain amount of judgment must be used to report the depth from where the sample came. Also, if shale sloughs off the side of the borehole and falls to the bottom during drilling, the alert geologist will understand that these chips of rock came from 400 feet down, not 1,200. The log of the raw samples may later be adjusted to match the wireline logs. The bit cuttings are rinsed, examined, packaged, labeled, and stored. The cuttings to be logged must be described thoroughly. Table 6.2 shows two examples, written in more complete form than the abbreviated style generally used.

TABLE 6.2

DESCRIPTION OF CUTTINGS

Characteristic	Limestone Sample	Sandstone Sample
Rock type	Oolitic grainstone	Lithic
Color	Brown	Buff to white
Texture	Medium to coarse	Fine to medium; angular
Secondary components or cements	Argillaceous	Slightly argillaceous
Fossils and accessories	Brachiopods and bryozoa; glauconitic	Mica
Sedimentary structures		
Porosity and oil shows	Good interparticle porosity; good oil stain; good cut fluorescence	Fair intergranular porosity; good oil stain; good cut fluorescence

Read the description of the limestone sample. It is an oolitic grainstone type (meaning oolites smaller than 2 mm in diameter), brown (by a color chart), medium (0.25 mm) to coarse (1 mm) in texture, with some clay (less than 10%), containing some brachiopod and bryozoa fossils and some glauconite (important because this green mineral forms in marine environments), with good (10% to 15%) interparticle porosity. The sample is stained with hydrocarbons and shows a significant amount of hydrocarbon in the *fluorescence test*. The test involves observing the sample under ultraviolet light. Hydrocarbons fluoresce under ultraviolet light. The sample may be "cut," or immersed in a solvent. If any hydrocarbons are present, they will dissolve and appear as streamers or streaks of color different from the solvent.

Each of the seven categories in the description requires one or more tests. Obviously the geologist spends a lot of time examining and testing each batch of cuttings during the course of a well sitting. A 10,000-foot well with samples every 10 feet means a thousand examinations over maybe a month's time. Usually the information gleaned from these examinations is plotted from the description in graphic form on a log strip chart. In figure 6.4 the two examples from table 6.2 are shown logged in. The logs are often done in color, which involves another symbology.

The log serves as source material for study by other geologists for years to come. The anticipated geologic column in table 6.1 was derived from logs. When the actual depth to the top of each formation is known, a new control point is established, and structural contour maps can be improved. The detailed lithology and fossil studies allow stratigraphic interpretations to be made, such as following facies changes across a basin or reconstructing a paleogeography.

Today, geologists routinely test samples from all porous formations as the well is drilled, because in that way oil is found. The giant Kelly-Snyder field was drilled through in 1947 but not discovered, because no geologist was on duty and no one else was

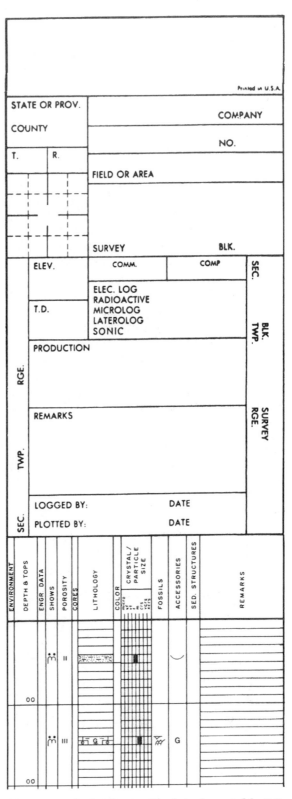

Figure 6.4. Strip chart with data from table 6.2 logged in

watching. The company was not interested in the Permian rocks; no oil was expected there. It was looking for the deeper Pennsylvanian rocks that produced in nearby counties. The Ellenburger formation proved dry; the well was abandoned, and the Kelly-Snyder field was not discovered until the following year.

The mud logger looks for oil in the mud and in the cuttings and runs a gas detector on the mud returns. Usually a portion of the mud coming out of the wellbore is routed through the detector and placed in a vacuum; the gases that come out of solution are tested for hydrocarbon content with a gas chromatograph or other sensitive instrument.

Sidewall samples. Special equipment may be used to take larger rock samples from known depths. Sidewall coring devices are lowered into the well on wireline or pipe to the depth of the formation to be sampled. One type uses explosive charges to fire hollow metal bullets into the sidewall (fig. 6.5); another type bores into the sidewall. A typical sample is about ¾ inch in diameter and maybe 1½ inches long. Like

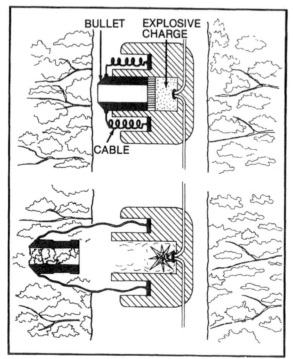

Figure 6.5. Sidewall coring operation

Figure 6.6. Core being retrieved from core barrel

bit cuttings, sidewall samples are logged, labeled, and stored for later examination.

Cores. To obtain even larger rock samples for geologic analysis, the rotary bit may be replaced with a doughnut-shaped coring bit. The coring bit makes hole like a rotary bit, except that a cylindrical core is produced through its open center. The core passes through the coring bit into a core barrel, which catches the sample and contains it if it breaks or crumbles (fig. 6.6). Some core barrels include a flexible plastic or foam rubber sleeve that not only protects the sample from damage but also prevents the loss of most fluids from the core. Cores may be of any length, but 30 to 90 feet is normal. Core sample diameters range in size from 1⅛ to 5 inches or larger.

The decision to core or not is usually made before the well is drilled, but the geologist at the well site determines when to stop drilling and begin coring. The geologist examines the core upon retrieval and logs it like other samples. The core is cut into sections, labeled, and packaged for shipment to the lab. Because of the expense, not all wells are cored. However, more money is spent on the gathering of basic data during the early part of exploration and field development than later, so cores are commonly taken in rank wildcats and early development wells.

The laboratory analysis of core samples is usually very detailed. Typically the core will be tested for porosity, permeability, and liquid saturations. Additional tests involve measurements of chloride content, capillary pressure, resistivity, relative permeability, and other characteristics. Clay content is evaluated through centrifuge or stain analysis. Thin slices are examined under a scanning electron microscope or other sophisticated instrument. Other tests may be conducted at later times.

Cores are taken from exploratory wells to obtain additional subsurface structural and stratigraphic information, to calibrate wireline logs, and to evaluate production characteristics, primarily by various measures of permeability. Nothing else provides the direct evidence that a core can. However, the core is usually not the conclusive factor in deciding whether to complete the well or not, but is rather one of many pieces of evidence.

All rock samples are available for later study by paleontologists, sedimentologists, and other specialists. Petrological studies, heavy mineral analyses, geochemical studies, and other work may be called for. In general, there is no reason to dispose of the original evidence. It may be needed later. The material may eventually be donated to educational institutions such as The University of Texas at Austin, Core Research Center of the Bureau of Economic Geology.

Wireline Well Logs

As if not busy enough with rock samples, the geologist must also help supervise wireline logging operations and inspect and sign for the logs. Wireline well logging involves lowering various sensors, encased in a tube called a *sonde*, into the well on a cable (fig. 6.7). Logging sondes have been designed to gather physical, electrical, radioactivity, and acoustic data from subsurface formations. Each type has many variations. The records that each produces are called logs.

The geologist notifies the logging company several hours in advance of the logging

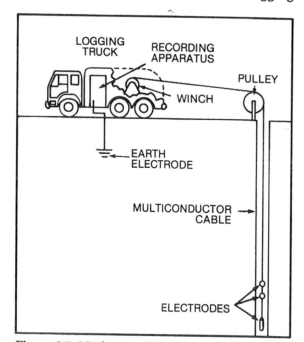

Figure 6.7. Mechanics of lowering a logging sonde into the wellbore

Figure 6.8. Logging sonde

operation so that a crew can show up on time with the proper equipment. The logging contractor is informed of various characteristics of the well, such as elevation at the rig floor, depths to casing points and total depth, casing sizes, hole conditions, and drilling mud properties. These features affect not only the running of the tools but also the reading of the logs.

The appropriate tools are assembled and lowered on a cable to the bottom of the well. As the sonde is reeled in, the signals transmitted up the cable are recorded on a strip log and usually as a digital record on magnetic tape. Copies are made for the engineers and the geologist on site. A computer may or may not be used for correc-

tions or calculations. Before the logging tools are put away (fig. 6.8), the geologist must make sure that the logs obtained are of acceptable quality.

Caliper logs. The diameter of the wellbore changes during drilling for a number of reasons. The drill bit tends to move from side to side as it penetrates rock layers. Some of the weaker layers may crumble or cave in after the bit has passed through. As the caliper logging sonde travels upward, arms or springs that press against the sides of the hole move in and out in response to the hole's narrowing and widening (fig. 6.9). The arm movements are immediately translated into electrical signals and sent to the surface to be logged. A

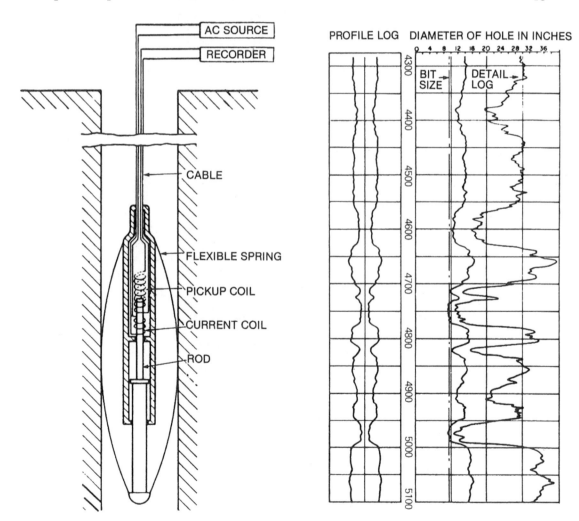

Figure 6.9. Caliper device and caliper log presentation

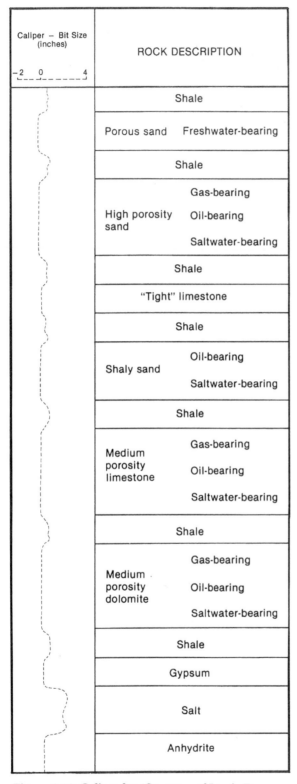

Caliper – Bit Size (inches) -2 0 4	ROCK DESCRIPTION	
	Shale	
	Porous sand	Freshwater-bearing
	Shale	
	High porosity sand	Gas-bearing
		Oil-bearing
		Saltwater-bearing
	Shale	
	"Tight" limestone	
	Shale	
	Shaly sand	Oil-bearing
		Saltwater-bearing
	Shale	
	Medium porosity limestone	Gas-bearing
		Oil-bearing
		Saltwater-bearing
	Shale	
	Medium porosity dolomite	Gas-bearing
		Oil-bearing
		Saltwater-bearing
	Shale	
	Gypsum	
	Salt	
	Anhydrite	

Figure 6.10. Caliper log (Courtesy of Lewis Raymer, Schlumberger Well Services)

typical response of the caliper log is shown in figure 6.10. Caliper data are used for planning well casing, cementing, and completion programs.

Dipmeter logs. The dipmeter sonde has three or four electrodes that contact the walls of the hole. Each of the electrodes sends a separate stream of data to the surface. Since the electrodes are level with one another, a formation that is not horizontal shows up on each of the logging curves at slightly different times. The order in which a formation shows up on the curves and the differences in time allow a dip angle calculation (fig. 6.11). Dipmeters are digitally recorded and computer-processed to determine dip angle.

Acoustic logs. A typical acoustic, or sonic, sonde contains a sound generator and two receivers, with the receivers spaced at different distances from the generator (fig. 6.12). As it is drawn up the mud-filled wellbore, the generator transmits sound

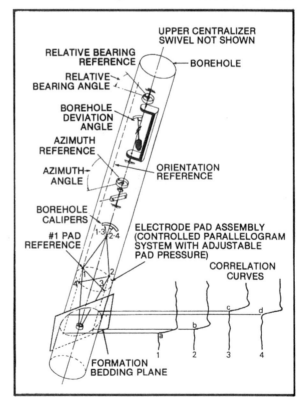

Figure 6.11. Schematic diagram of a four-armed diplog (Courtesy of Dresser Atlas, Dresser Industries, Inc.)

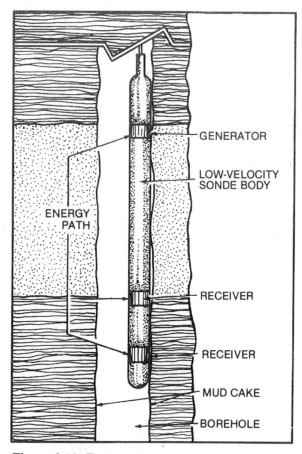

Figure 6.12. Two-receiver acoustic logging sonde

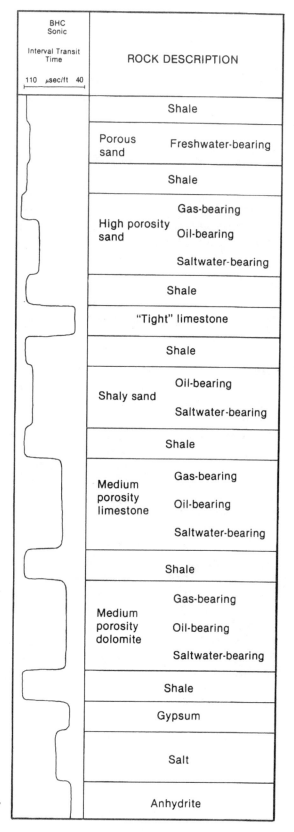

waves into the formation to be picked up by the receivers. The travel time between source and receiver is used to calculate the speed of sound in the formation. The second receiver helps to correct for borehole irregularities. Acoustic velocity varies with rock type and porosity (fig. 6.13). Generally, the denser a formation is, the faster it conducts sound. Acoustic waves travel slowly through fluids and gases, so travel time is longer in high-porosity than in low-porosity rock. The travel time can be used to estimate the porosity.

Figure 6.13. Sonic log (Courtesy of Lewis Raymer, Schlumberger Well Services)

Spontaneous potential logs. When two liquids of differing salinity are placed together, ions from the saltier liquid tend to move into the less salty one until they are equivalent solutions. However, if a physical barrier of low permeability separates the two solutions, the difference in salinity creates an electric potential, and a weak current flows toward the solution with lower salt concentration. For example, a potential for current to flow between the formation waters is created at the boundary between an impermeable shale and a permeable sandstone. Around the wellbore, currents are produced by differences in the salinity of drilling mud and of formation water. The electrode in the sonde of the spontaneous potential (SP) log senses the current flow relative to an electrode that is grounded. The response in millivolts is logged against well depth.

The primary use of the SP log is to determine which strata are porous and which are not (fig. 6.14). The SP curve moves to the right when the sonde is adjacent to impermeable, electrically conductive beds such as shales. A shale baseline can often be established down the right margin of the plot. This line may shift because of changes in salinities with depth. Deflection of the plot to the left, off the baseline, indicates a porous and permeable bed such as a sandstone or a carbonate.

The magnitude of the deflection indicates a difference in salinity between the drilling fluid and the formation water and is also influenced by the thickness of the beds, the shaliness of the permeable beds, and other factors. Correcting for these factors and knowing the salinity of the drilling fluid and the temperature in the formation allow calculation of the resistivity of the water in the formation of interest.

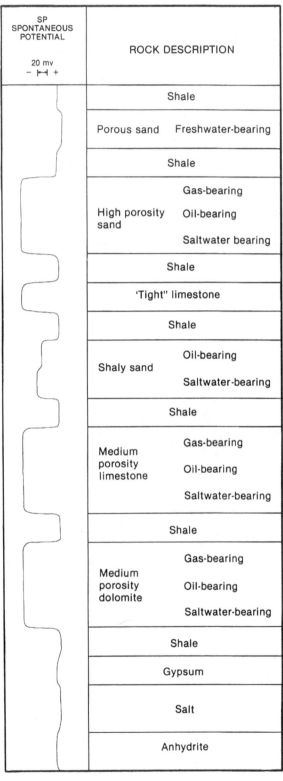

Figure 6.14. Spontaneous potential log (Courtesy of Lewis Raymer, Schlumberger Well Services)

Resistivity logs. While SP logs are measurements of a formation's natural electrical characteristics, resistivity logs record a reaction to electric current from outside sources. A mineral/rock matrix usually shows high electrical resistance. The resistance of water depends upon the concentration of dissolved salts in it. As the salinity increases, resistance decreases. Resistance to current flow through a few feet of subsurface formation depends on a number of things, such as total porosity, pore geometry, salinity of the formation water, and the amount of oil or gas present.

It is important to remember that the drilling fluid will invade a permeable zone. Safe drilling practices call for the adjustment of drilling fluid density to create pressure that is greater than or equal to the formation pressure. Otherwise fluids from the formation may enter the wellbore and even lead to a dangerous blowout of the well. Electric logs are very sensitive to the effects of the invasion of the formation by drilling fluids (fig. 6.15). The electrical properties of the invaded zone mask those of the unaffected formation and complicate the reading of the logs. The purpose of resistivity logs is to determine resistivity through the invaded and the undisturbed formation.

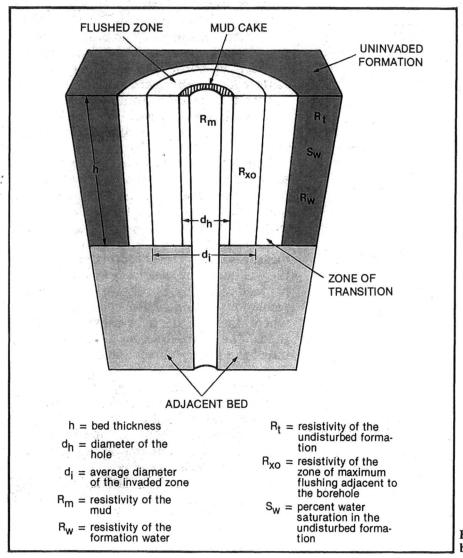

FLUSHED ZONE   MUD CAKE

UNINVADED FORMATION

R_m

R_t

S_w

R_{xo}

R_w

h

d_h

d_i

ZONE OF TRANSITION

ADJACENT BED

h = bed thickness

d_h = diameter of the hole

d_i = average diameter of the invaded zone

R_m = resistivity of the mud

R_w = resistivity of the formation water

R_t = resistivity of the undisturbed formation

R_{xo} = resistivity of the zone of maximum flushing adjacent to the borehole

S_w = percent water saturation in the undisturbed formation

Figure 6.15. Diagram of borehole conditions

A set of electrodes in the resistivity tool produces an electric current in the rock

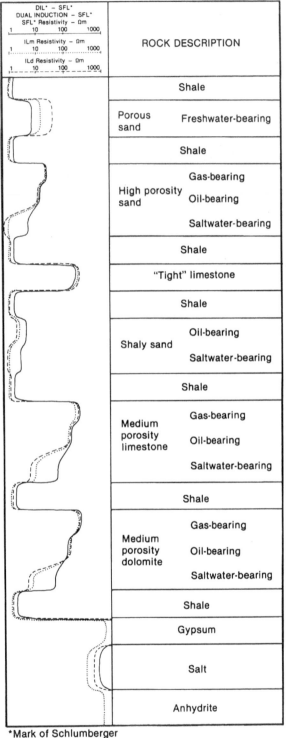

*Mark of Schlumberger

Figure 6.16. Resistivity logs (Courtesy of Lewis Raymer, Schlumberger Well Services)

around the wellbore. A separate set of electrodes in another part of the sonde picks up the current that has traveled through the formation. After flowing through a relatively resistive formation, the current is weaker than that which would have flowed through a less resistive one. The current measurements are plotted against well depth as resistivity (fig. 6.16) or as conductivity, which is the reciprocal of resistivity.

The *induction log* sonde uses a sonde that contains two sets of wire coils, one to transmit and one to receive. Current in the transmitter coils produces a magnetic field. A current is induced in the formation, which in turn produces its own magnetic field, causing a current to flow in the receiver coils of the sonde. The strength of the current in the receiver coils is directly proportional to formation conductivity and inversely proportional to resistivity. An advantage over other resistivity logs is that this sonde does not require conductive wellbore fluids for successful operation, so it can be used in empty wellbores or in those filled with oil-base mud.

Several resistivity and induction tools are used today. The differences lie mainly in the way that the current is focused around the wellbore by the spacing of the electrodes on the tool. The *short normal log,* rarely used in oil exploration today, typically measures about 3 feet of formation out from the wellbore. The *spherically focused log* does not read as deeply, but shows better vertical resolution. The *dual induction log* is used where there is deep invasion of drilling fluids. The *lateral focus log* and the *guard log* direct the survey current laterally in a thin disc or ring shape. As the sonde is raised in the wellbore during logging, formations only inches thick may show up clearly on the log. Lateral focus logs and guard logs are particularly useful when the hole is full of saltwater drilling mud that would adversely affect other tools.

While the lateral focus and induction logs respond to resistivity out to several feet from the wellbore, the *microresistivity* log responds to only a few inches. Knowing the characteristics of the invaded zone allows correction for their effects on other logs. The sonde has an insulated pad that is held firmly against the wellbore sidewall to seal off a small area and minimize interference from wellbore fluid resistivity. Electrodes in the pad respond to resistivity at two distances: a shallow one of perhaps 1 inch, usually no more than mud cake thickness, and a slightly deeper one of several inches. The two resistivity curves are plotted on the same log. When the pad is in contact with relatively impermeable formations, such as shale, there is no mud cake, and the two curves are almost identical. But when the pad is pressed into mud cake, the curves separate and show the effects of invasion by mud filtrates.

Radioactivity logs. Radioactivity logging can measure both natural and induced radioactive characteristics of formations. The radiation levels involved are usually very low but nevertheless detectable and useful to the geologist. Radioactivity logs must be run more slowly than electric logs because they are measuring statistical events.

Scattered throughout most sediments are three radioactive elements—potassium 40, thorium, and uranium. The *gamma ray log* (fig. 6.17) is a record of the natural radioactivity around the wellbore. Shales generally produce higher levels of gamma radiation than sandstones or limestones and can be detected and studied with the gamma ray tools. In holes using salty drilling fluid the gamma ray log is more effective than SP tools for distinguishing shales.

The *neutron log* bombards the rock around the wellbore with radiation and records the amount of radiation that is not absorbed. The sonde contains a neutron source and two receivers. Neutrons are slowed by collision with particles of similar mass. The hydrogen nucleus has about the same mass as a neutron, so the tool responds mainly to the hydrogen ion content of the formation. The hydrogen ion content can be used to estimate porosity. Oil and water have about the same hydrogen ion content by volume, but in a zone containing dry gas the neutron log shows less than the actual porosity because the gas has a lower hydrogen ion

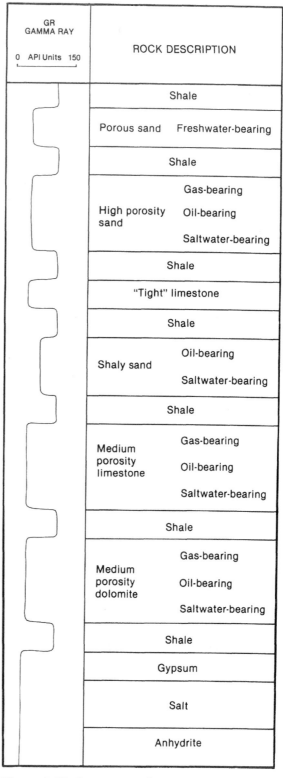

Figure 6.17. Gamma ray log (Courtesy of Lewis Raymer, Schlumberger Well Services)

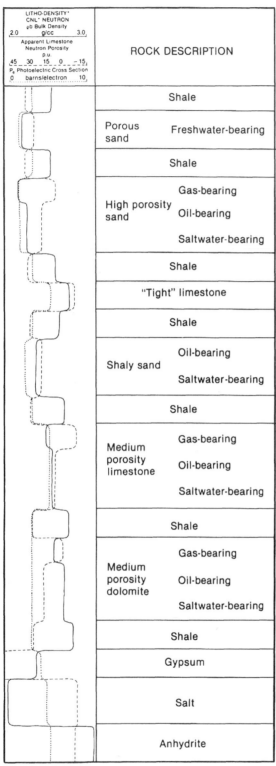

LITHO-DENSITY* CNL* NEUTRON φb Bulk Density 2.0 g/cc 3.0 Apparent Limestone Neutron Porosity p.u. 45 30 15 0 -15 Pe Photoelectric Cross Section 0 barns/electron 10	ROCK DESCRIPTION	
	Shale	
	Porous sand	Freshwater-bearing
	Shale	
	High porosity sand	Gas-bearing
		Oil-bearing
		Saltwater-bearing
	Shale	
	"Tight" limestone	
	Shale	
	Shaly sand	Oil-bearing
		Saltwater-bearing
	Shale	
	Medium porosity limestone	Gas-bearing
		Oil-bearing
		Saltwater-bearing
	Shale	
	Medium porosity dolomite	Gas-bearing
		Oil-bearing
		Saltwater-bearing
	Shale	
	Gypsum	
	Salt	
	Anhydrite	

*Mark of Schlumberger

Figure 6.18. Radioactivity logs (Courtesy of Lewis Raymer, Schlumberger Well Services)

content by volume than water or oil. The neutron log is a device to indicate porosity. The tool is calibrated to a specific type of rock, conventionally showing a limestone porosity plot (fig. 6.18). At points where the tool has been next to limestone, the porosity values can be read directly off the log. The porosities of the other lithologies must be derived from the values on the log.

The *density log* radiates the formation with gamma rays, and receivers detect those not scattered by the electron density of the rock matrix. If the density of the minerals and the fluids content of a rock are known, porosity can be calculated from the density log. The density log is particularly useful in shaly sands and vuggy limes where sonic logs can be misleading. The density log in figure 6.18 shows how the line moves to the right for higher densities and lower porosities, and to the left for lower densities and higher porosities..

A new type of log, the *photoelectric absorption cross section,* or P_e, is run to analyze the mineral components of formations. The tool measures energy levels of incident gamma rays to produce a log of index numbers. The numbers are correlated with mineral densities to find out the proportions of mineral types and the porosity of the rock. If the chemical makeup of the formation is known, the P_e curve alone can be read to interpret the lithology (fig. 6.18). It can be used with the density log to determine porosities, or used with the density and neutron logs to analyze more complex lithologies and porosity.

Log Interpretation

The types of logs to be run in a particular hole are usually selected before drilling, although hole conditions may necessitate a change of plans. These logs may be used by different geologists and engineers for different purposes over many years. For completing an exploratory well, the log suite, or group of logs run together, needs to locate zones of porosity and indicate their water saturation. Examine the logs shown in figure 6.19. The columns to the left of the

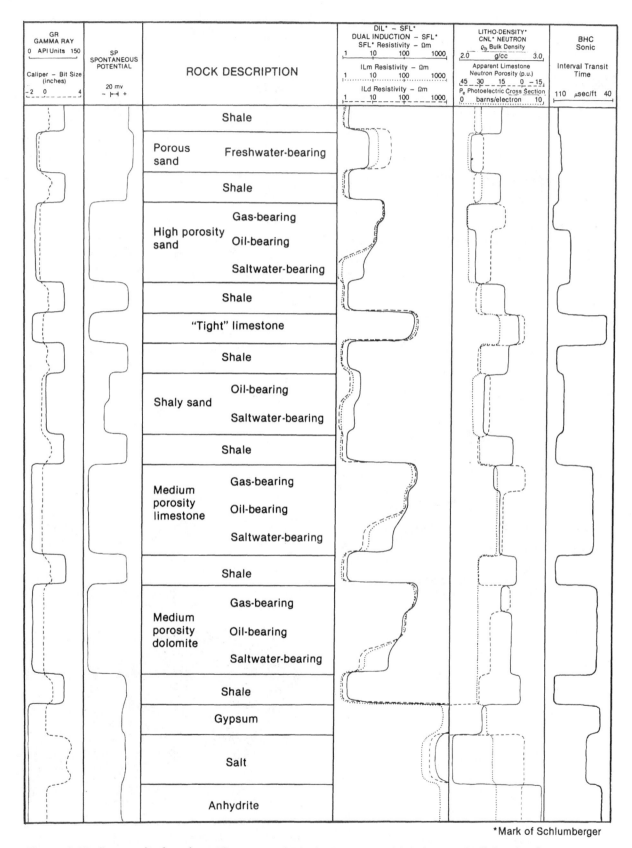

Figure 6.19. Composite log chart (Courtesy of Lewis Raymer, Schlumberger Well Services)

lithology description show three logs that indicate shale conditions and the existence of permeable zones. The columns to the right of the lithology description show seven logs that can be interpreted for porosity and water saturation values. All of the logs shown in the figure would probably not be run together, but the combination of logs selected for the run will include one or more of the types from each side. Each type of logging tool has limitations that depend upon hole conditions, so the geologist must match the capabilities of the tools with the situation in order to select the proper suite.

To figure the percentage of oil or gas saturation of the volume around the wellbore, the amount of water saturation is subtracted from the total pore space. The calculations of water saturation, developed by G.E. Archie in the 1940s, require values for three factors:

1. resistivity of the water in the formation of interest, R_w, usually calculated from the SP log;
2. resistivity of the undisturbed formation, R_t, usually calculated from an induction log or lateral focus log; and
3. formation resistivity factor, F, calculated from an acoustic log, a density log, a neutron log, or a core sample.

The relationships of the factors vary, but for a clean sandstone,

$$S_w{}^2 = \frac{F \times R_w}{R_t}$$

where S_w is the water saturation.

Assume, for example, a clean sandstone where

R_w = 0.095
R_t = 15
F = 20.83.

What is the water saturation?

$$S_w{}^2 = \frac{20.83 \times 0.095}{15} = 0.13$$

$$S_w = 0.36 = 36\%.$$

What is the saturation of oil and gas?

$$S_h = 1 - 0.36 = 0.64 = 64\%.$$

A large estimate of oil and gas in place does not guarantee that development is going to be a success, however. The commercial value also depends on other factors, such as rate of production.

FORMATION TESTING

How fast and how long a well will produce is predicted from calculations made with formation test data. The two types of formation tests commonly used on exploratory wells are the wireline formation test and the drill stem test. The wireline test data are recorded on magnetic tape and on paper or photographic film. Drill stem test data are recorded on metal pressure charts (fig. 6.20), inscribed by pressure recorders during a drill stem test. Some drill stem tests are now being logged through the use of electronic pressure recorders connected to surface equipment by conductor line. In all of these cases, the formation test data include recorded pressures. Formation testing is usually the responsibility of a petroleum engineer, who plans and supervises the test at the well site.

A sample of formation fluid is often taken during the tests. The sample usually amounts to a few gallons in the case of a wireline test but may be hundreds of gallons in a drill stem test. A laboratory study reveals characteristics of the petroleum that

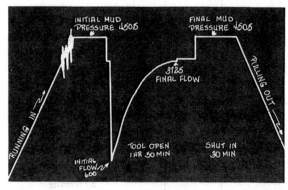

Figure 6.20. Bottomhole pressure record of a drill stem test

affect the type of production equipment to be installed at the well.

Formation Test Data

Pressure data consist of static and flowing bottomhole pressures. *Flowing bottomhole pressure* is the pressure at the producing formation face when the well is allowed to flow. *Static bottomhole pressure* is the maximum reading when there is no flow up the wellbore. The procedure involves isolating the formation with packers, allowing fluid to flow, shutting in the well, and allowing the pressure to build up again. When fluid flows, the pressure drops, and this portion of the record is called the *drawdown*. The record of the time during which static pressure slowly increases is termed the *buildup*. The ability of the well to produce its fluids is shown by the difference between static and flowing bottomhole pressures.

Predicting Production Behavior

Deliverability plots. A deliverability plot compares flowing bottomhole pressure with production in barrels of oil per day. The main purpose of deliverability plotting is to find the most efficient flow rate for the well. Some wells produce most efficiently if they are restricted to low flow rates; other wells must be produced quickly at high rates to obtain maximum production during well life.

Pressure buildup plots. In the early 1950s, a petroleum engineer named D.R. Horner developed the original techniques of pressure buildup analysis. The techniques involve reading pressures from the buildup portions of a formation test, using them to solve a simple logarithmic equation, and plotting the answers in a curve. The plotted curve, in turn, contains a number of mathematical values useful in calculations important to reservoir engineering and production management during the life of the well. These include equations for average effective reservoir permeability, well drainage

area, flow irregularities near the wellbore, and formation damage from poor drilling practices.

COMPLETING THE DISCOVERY

Before a decision can be made to case and complete a well, a reservoir must be located, its fluid content evaluated, and a fairly clear idea of the well's future production behavior developed. Only then can the oil company decide whether to spend more money on the well. A well that has good potential for sustaining commercial production will be completed.

Evaluation

By the time the well has reached total depth, the geologist may already have a pretty good idea of the chances for production. First, all shows of hydrocarbons noted during drilling are evaluated. A show does not necessarily indicate commercially valuable reserves. The quantity of oil and gas may be insufficient, or the rock may be inadequate to serve as a viable reservoir. Second, all well logs and test data are evaluated to find potentially productive zones. Because the drilling mud flushes the rock ahead of the drill bit, a productive zone may not yield a mud log show of hydrocarbons. Rock bodies with good porosity and permeability, evidenced by drilling breaks, circulation losses, or other means, are evaluated on logs. Additional tests may be called for in zones where hydrocarbons are suspected to exist. Proper evaluation of permeable zones and hydrocarbons will enable the decision to be made about whether to complete the well or not. The geologist will examine all of the evidence, correct the lithologic log (fig. 6.21), and prepare a recommendation, following company procedures.

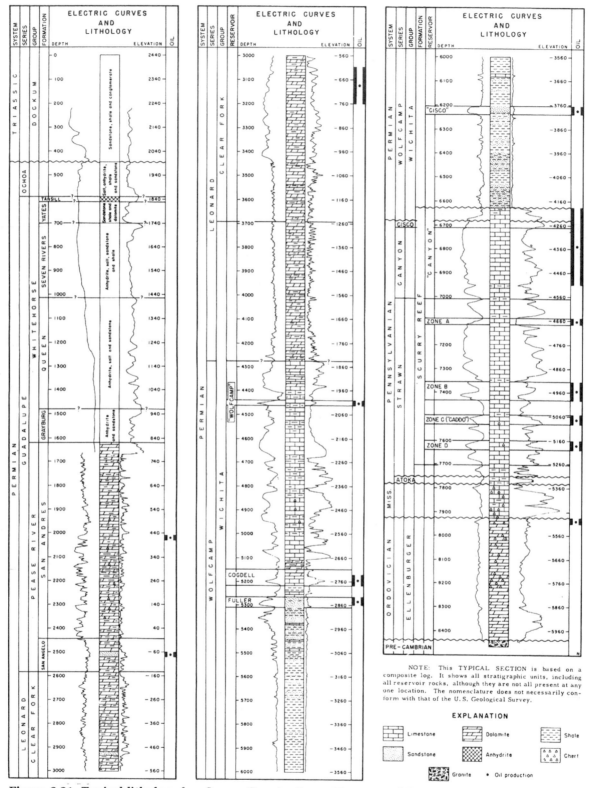

Figure 6.21. Typical lithology log, Scurry County, Texas (Courtesy of Bureau of Economic Geology, The University of Texas at Austin)

Completion

The completion technique is chosen by the engineers. In a new region, the engineers will consult extensively with the geologist about formation characteristics. In a well-known play, completion practices may be standardized. The completion engineer who selects the particular technique to be used on a given well must consider the cost; the possibility that the formation has been damaged by infiltration of the drilling mud; the possibility that well stimulation treatments will be needed; the type of reservoir drive; locations of oil-water and gas-oil contacts; and future needs for artificial lift, ser-

vice and workover projects, and secondary recovery operations.

The geologist's input during analysis for completion helps determine the specifics of a completion. If bits of clay are going to flow into the well or if sand or corrosive fluids will be produced, the engineers need to know as far in advance of completion as possible. The geologist's advice may be sought on the vertical and radial positioning of perforations or on a stimulation treatment.

Offshore, the situation is somewhat different. Not until sufficient information comes in does the development phase begin (fig. 6.22). Exploratory wells are normally

Figure 6.22.
Production
platform

plugged and abandoned and their data used to select completion techniques for production wells yet to be drilled. Completion is the same as that onshore, inasmuch as it conforms to the reservoir and not to the surface.

By far the most common type of completion calls for installing tubing with a packer and perforating the casing and cement to establish a flow path from the formation to the surface (fig. 6.23). First, production casing is cemented in place. This expensive set of tubular goods will enable control of the well to be exercised through a variety of operations and procedures. After a wellhead has been installed, a completion rig (fig 6.24) is brought in to install the tubing and packer, tubing head and hangers, and

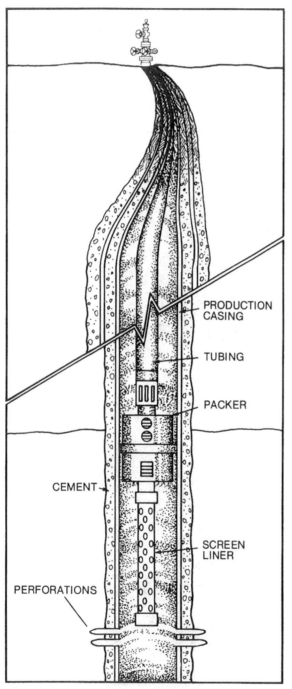

Figure 6.23. Conventional completion

Figure 6.24. Completion rig

Christmas tree (fig. 6.25). Sometimes tubing is run and set prior to perforating, and sometimes the casing is perforated and then tubing is run (fig. 6.26). Tubing is used because it can be replaced, repaired, or fitted with a sucker rod pump, and it provides desirable control over the well. Production casing cannot be easily replaced or repaired and is not easily fitted with pumping devices.

Perforations are needed to penetrate through the steel production casing and cement and into the producing formation to allow fluid to pass from the rock into the wellbore. Jet perforating is used today, typically shooting four to eight holes per vertical foot. A cone-shaped charge explodes, and the jet stream exerts enough force to reach as far as 2 feet into the formation (fig. 6.27). After the "perfing," several barrels of fluid are produced to clean up the well and remove debris; then the well is

Figure 6.26. Tubing run

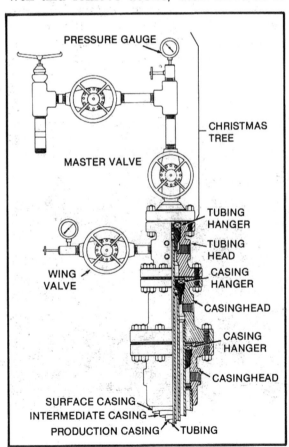

Figure 6.25. Christmas tree

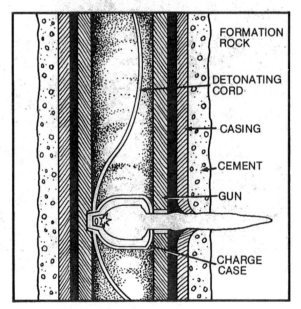

Figure 6.27. Perforating

tested and placed on stream. Swabbing is an old technique, still used, to lower the pressure in the tubing and bring in the oil. A pump may be needed to sustain oil production.

Within the limited goal of completing a well are a surprising number of variations in technique. A well may be completed "barefoot," or open-hole. In thick, massive, well-consolidated carbonates, the production casing may be set above the pay zone. In this case the wellbore through the pay zone is literally an open hole. A screen liner and gravel pack may be needed to keep sand out of the wellbore. Multiple pay zones require separate flow paths to the surface. Wells can be completed in a number of configurations, with single-string dual completions, parallel string completions, miniaturized multiple completions, and so on.

Many wells need stimulation in order to come in profitably. Two invasive methods of stimulation are hydraulic fracturing and acidizing. The high-pressure injection of fluids (fig. 6.28) such as water, oil, or acid will fracture the rock. Proppants and spacer

materials prevent the channels from collapsing afterward. Matrix acidizing occurs when acid is pumped down at pressures below those needed to fracture the formation. The acid etches the natural permeability paths and may improve production.

EXPLORATION SUCCESS

A pay zone is the section of wellbore that penetrates a commercially producible reservoir. Pay is not usually found, especially in wildcat wells, so most wells are plugged and abandoned before completion. The geologist adds the information obtained to what is already known and starts over. The understanding gained by drilling helps with future operations. The operators would not be able to continue in business, however, if the bonanza wells did not pay out enough to cover the expenses of all the stinkers and dusters.

Costs are always an important factor in planning. The expected life of the project and the place of the well in development of the field affect the economics. Costs include

Figure 6.28. Fracturing

present and future outlays, which have to be balanced with the expected value of production and the rate of production, not only of the exploration wells but also of the whole field. Success can be measured in many ways, but for the geologist, the confirmation of months of work by production must be tremendously satisfying.

Exploration is an expensive and frustrating endeavor. The geologist knows the extent to which geology contributed in finding the oil and gas, and should be prepared to defend the role of science in the venture, as well as to learn from the mistakes. Many oil companies do not even attempt to find new fields. Nonetheless, a successful exploration venture is prerequisite to the development of the petroleum reserves so in demand today. The wellsite geologist is still key to the modern exploration picture.

FIELD DEVELOPMENT

EXPANDING AND MAINTAINING PRODUCTIVITY

Developing a field requires a solid grasp of regional geology and reliable well data. It is desirable to have a single geologist or a single team work on the development because so much depends on the previous history of each drilled well. If the development geologist is not the same person as the exploration geologist, the two will consult extensively. Likewise, the development team will work with the exploration department, sharing insight and plans. While the exploration program may call for perhaps three wells to locate the structural crest along a simple anticlinal structure, the development program moves to account for the complexities of the reservoir. Productivity within the structure will vary.

Selecting Well Sites

Although only the surface can be seen (fig. 7.1), selecting sites for the wells succeeding the discovery well requires the geologist to visualize the reservoir. Generally the position expected to be structurally highest will be drilled first. This and the next few well sites are commonly picked on the basis of the seismic picture. As wells are drilled, formation velocity surveys of the reflecting surfaces are made downhole.

These enable the seismic maps to be expressed more accurately in terms of depth, and therefore a better picture of the structural high can be generated. As subsurface contour maps are developed by the geologist, they replace the seismic picture for decision-making purposes.

The next well will be drilled at the next most favorable location. Usually this means an offset well on an adjacent drilling unit according to the spacing requirements for the field. The direction from the previous well depends upon the reservoir. If a water drive reservoir is expected, the next well may be drilled updip of the previous well. If the reservoir has a gas cap, the next well may be drilled downdip.

From the structural high point below the gas-oil contact, development will proceed down the flanks of an anticline to determine the productive limits of the new field. In figure 7.2 the structure is shown being drilled on alternating sides of the discovery, first along the axis of the anticline and then at right angles.

Sometimes it is important to establish the productive limits of the reservoir early. The offset can be drilled later, and its site is skipped over in the search for the edge of the reservoir. Wells drilled beyond adjacent sites are termed *step-outs*. Reasons for drilling step-out wells, other than determining

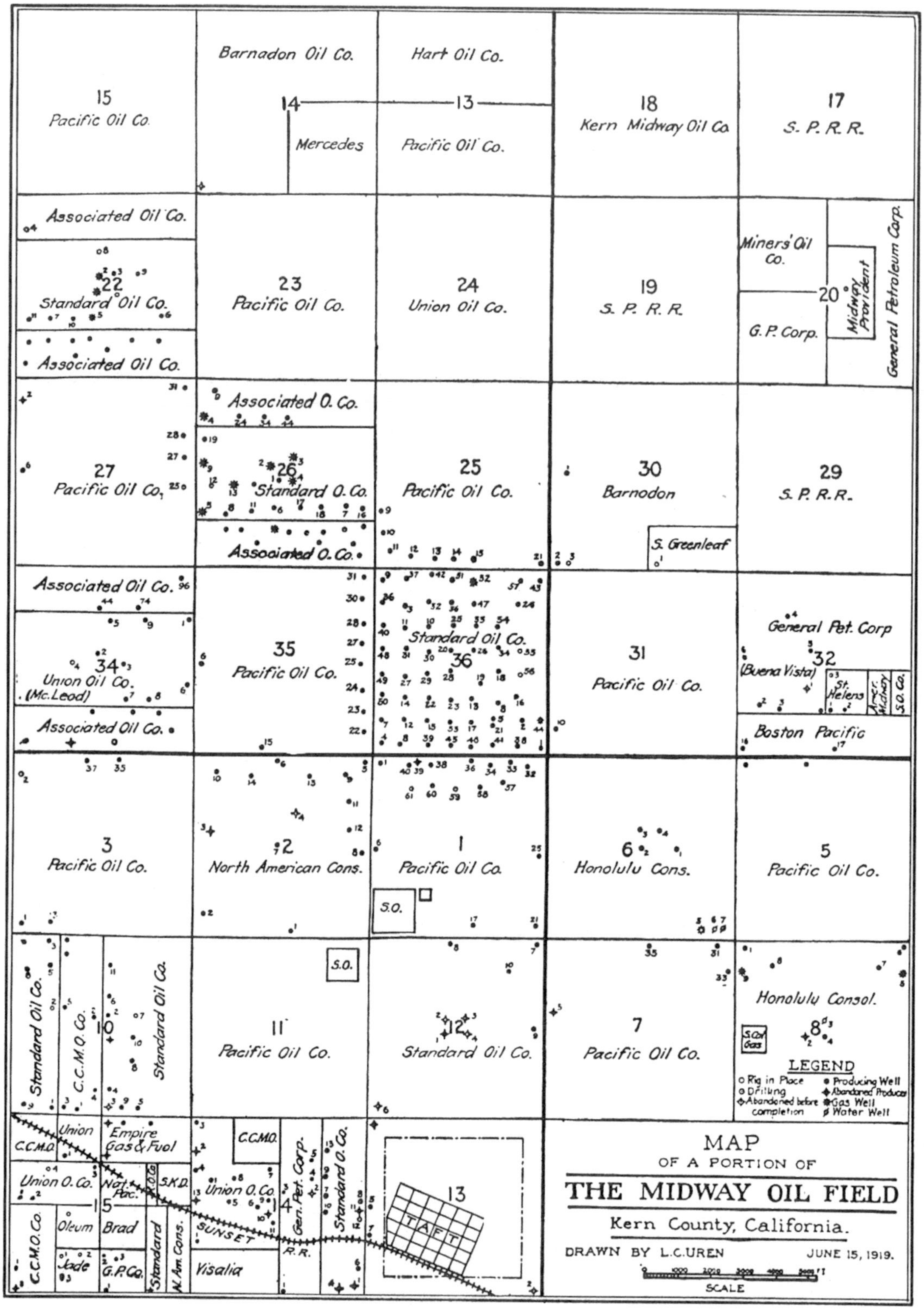

Figure 7.1. Land map of field development (From *Petroleum Production Engineering: Oil Field Development*, 2d ed., by Lester C. Uren. Copyright 1934 by McGraw-Hill Publishing Company. Reprinted by permission of the publisher.)

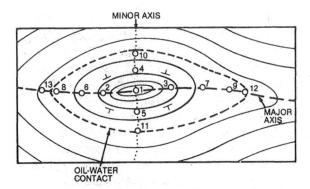

Figure 7.2. Structure contour map showing the order of drilling test wells to determine the productive limits of a new field (From *Petroleum Production Engineering: Oil Field Development*, 2d ed., by Lester C. Uren. Copyright 1934 by McGraw-Hill Publishing Company. Reprinted by permission of the publisher.)

the reservoir edge, may involve the need to prove up acreage before leases expire or needs that concern production or transportation.

In the United States, leases for an entire reservoir are rarely held by one company. Competing producers will have acquired rights in the area long before a discovery well is drilled. It is customary, in consideration of drainage across property lines, to drill along the boundary lines of a leased property before developing the interior. The operator who first brings his property into full development will produce more of his neighbors' oil than they of his.

Notice in figure 7.3 that pressure differential causes migration (from high to low pressure), and that the pressure gradient between the wells is determined in part by the production rates of the wells. If neighboring leases each have sixteen wells (fig. 7.3, case 3), and lease *A* produces at 100 bbl/day/well and lease *B* at 50 bbl/day/well, then lease *A* will withdraw 1,600 bbl and lease *B*, 800 bbl. Pressure will be lower at lease *A*, and, assuming good permeability between the two leases, migration will occur, with flow from *B* to *A*.

The competitive situation requires careful analysis (fig. 7.4). Overall efficiency can often be improved by cooperation among

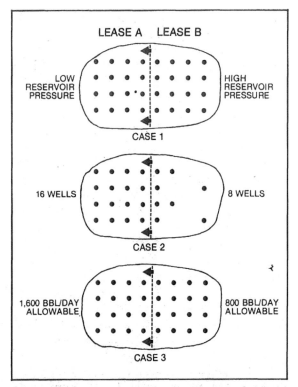

Figure 7.3. Indications of migration in the field across lease boundaries, with arrows showing direction of oil flow (From *Elements of Petroleum Reservoirs*, 1960. Copyright 1960 SPE-AIME.)

Figure 7.4. Overcrowding in Spindletop oil field, Texas, 1920

the operators. Agreements between producers on well sites along lease lines, regulations on well spacing, production allowables, prorationing, field-wide rules, and unitization contracts are commonly used devices that try to provide equitable conditions for competition and maximum ultimate recovery of the resource. The operator generally has both expressed and implied covenants with the lessor to develop the resource as completely as possible. The lessor will be helped most by the production policies that also benefit the operator most.

Recovery Efficiency

The total amount of oil and gas recovered from a reservoir depends on engineering practices and economic factors as well as on essential geology. Although the rate of production will surely have an impact on total recovery, the most efficient rate of production cannot easily be determined. Furthermore, the oil can be recovered only once. Analyses of rock and fluid properties, pressure histories, and production histories can indicate reservoir drive performance. In a water drive reservoir it is desirable for the interface zone, or water front, to advance smoothly and evenly in order to push out the most oil. Coning and fingering of water in advance of the front may call for adjusting production rates of wells in that area. In a gas-cap drive reservoir, the gas cap should not be allowed to shrink, possibly decreasing available drive energy and increasing the amount of unproducible, residual oil. Production rates of wells in a gas reservoir affect the amount of condensate produced over time. As pressure drops, condensates can form in the reservoir, making their recovery problematic.

The order of drilling influences ultimate recovery. Understanding the reservoir is necessary for wise investment. To best exploit the resources of a field, plans are made to drill between known producing wells (infill drilling) with information from years of production. Strategic infill drilling programs recognize that uniformly spaced wells are not always the best means of producing a reservoir that shows significant geologic variations and compartmentalization.

The rate of drilling also influences ultimate recovery. For example, the field shown in figure 7.5 would produce 2,210,000 barrels over 15 years according to hypothetical plan A, and 2,085,000 barrels over 15 years according to plan B. Average annual production would be 147,300 bbl/year for plan A versus 139,000 bbl/year for plan B.

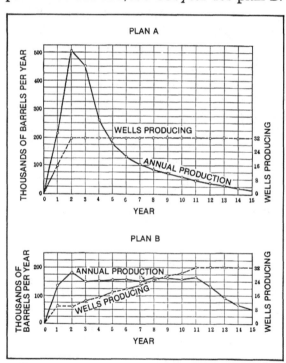

Figure 7.5. Estimated production of an oil property with rapid development, plan A, and with slow development planned to maintain uniform output, plan B. (From *Petroleum Production Engineering: Oil Field Development,* 2d ed., by Lester C. Uren. Copyright 1934 by McGraw-Hill Publishing Company. Reprinted by permission of the publisher.)

Note that in each case wells are assumed to be so spaced that they have equivalent initial productions and decline rates.

Which is the more successful plan is not immediately apparent, because the timing of the rate of return differs, as does the rate of investment. Nonetheless, the plan of rapid development produces more per year and more in total than the slower plan of development.

The optimum producing rate for a single well is one that allows production at the highest possible rate but still below the rate at which total recovery would be ultimately reduced. Conventional wisdom holds that a reservoir produced too rapidly will deplete the energy needed to drive oil or gas to the surface. To prevent waste of this type, state regulatory agencies control the production rates of wells. In Texas the Railroad Commission has established production allowables according to depth and well density, published in its *Rules and Regulations*. For example, a 5,000- to 6,000-foot well on 40-acre spacing is allowed 102 barrels per day, according to rules 45 and 52. The production flow path will be engineered for maximum efficiency within the rate set by the regulatory allowable.

Plugging and Abandonment

Individual wells that have outlived their usefulness are plugged and abandoned. Plugging and abandoning a well indicate that for economic reasons the oil and gas supply is exhausted. However, every well should be evaluated for the possibilities of pay zones either deeper or shallower than the zone of current production. If shallower possibilities exist, the well can be plugged back and perforated higher up in the casing. If deeper possibilities exist, the perforations can be filled up by squeeze cementing, and the well drilled deeper.

The abandonment procedure is spelled out by regulatory agencies that have responsibility for protecting the public from dangerous holes. In Texas the Railroad Commission requires a plugging report (fig. 7.6) to be submitted and plugging operations to begin within 90 days after production operations have ceased. The surface or conductor casing must be left in the ground, 100 feet of cement placed immediately above the perforated zone, fluid placed above the cement, and a 10-foot plug of cement placed at the top of the well. Additional cementing may be required by the district director of the commission. Occasionally wells are abandoned, and no one can be found to take responsibility, so the state must plug the well.

Fields are not abandoned in the sense that individual wells are, for even under favorable conditions of production, a great deal of the original resource still remains in the reservoir. Many fields have withered from neglect because of economic conditions. Sometimes a field may be retained by drastically cutting expenses. Plugging old wells, shutting in gassy oilwells, and maintaining leases by unitizing and producing from a minimal number of wells are possible ways to retain marginally economic fields. Then, if the relative price of oil increases and economics favors greater investment in the field, newer technologies can be tried. Efficiency is the key here, the assumption being that as each company tries to minimize marginal cost and maximize marginal revenue, the overall production will be at its most efficient point.

Today when a company can see no way to hold on to a field, it is apt to look for a buyer who might be in a position to make some profit out of any wells remaining. A producer who is not paying windfall profit tax or who can still get a depletion allowance might be in such a position.

RAILROAD COMMISSION OF TEXAS
Oil and Gas Division

Notice of Intention to Plug and Abandon

Operators must comply with RRC plugging procedures as outlined on the reverse side.

Type or print only

1. Operator's Name and Address (Exactly as shown on Form P-5, Organization Report)	3. RRC District No.	4. County of Well Site
	5. API No. 42-	6. Drilling Permit No.
7. Rule 37 Case No.	8. Oil Lease No. or Gas Well ID No.	9. Well No.

2. RRC Operator Number _____

10. Field Name (Exactly as shown on RRC records)	11. Lease Name

12. Location
● Section No. _____ Block No. _____ Survey _____ No. _____ Abstract No. A-

● Distance (in miles) and direction from a nearby town in this county (name the town). _____

13. Type of well
1 - oil   3 - disposal   5 - other (specify) _____
2 - gas   4 - injection   Enter appropriate no. in box ▶ ☐

14. Type of completion
Single ☐   Multiple ☐

15. Total depth

16. Usable-quality water strata (as determined by Texas Dept. of Water Resources) occur to a
depth of _____ feet and in deeper strata from _____ to _____ feet; and from _____ to _____ feet

17. ● If there are wells in this area which are producing from or have produced from a shallower zone, state depth of zone _____
● If there are wells into which salt water is being or has been disposed of into a shallower zone, state depth of zone _____

18. Casing record (list all casing in well)

Size	Depth	Cement (sacks)	Drilled hole size	Top of cement (feet)	Top of cement determined by: Temper. Survey	Calculated	Cement bond log	Anticipated casing recovery (feet)
_____ set @	_____ w/	_____	_____	_____	☐	☐	☐	_____
_____ set @	_____ w/	_____	_____	_____	☐	☐	☐	_____
_____ set @	_____ w/	_____	_____	_____	☐	☐	☐	_____
_____ set @	_____ w/	_____	_____	_____	☐	☐	☐	_____
_____ set @	_____ w/	_____	_____	_____	☐	☐	☐	_____

19. Has notice of Intent to plug been filed previously for this well?
☐ Yes ___/___/___ mo. day yr.   ☐ No

20. Plugging proposal (List all bridge and cement plugs. Load the hole with at least 9.5 lbs. per gallon mud.)

21. Record of perforated intervals or open hole

Perforations	Open	Plugged	Plugging method
_____	☐	☐	_____
_____	☐	☐	_____
_____	☐	☐	_____
Open Hole	☐	☐	
_____	☐	☐	

No. of sacks   Depth in feet (top & bottom)
1. _____   _____
2. _____   _____
3. _____   _____
4. _____   _____
5. _____   _____
6. _____   _____
7. _____   _____
8. _____   _____

22. Name and address of cementing company or contractor

23. Anticipated plugging date for this well is: ___/___/___ mo. day yr.

Typed or printed name of operator's representative

Title of person

Telephone: Area Code Number   Date: ___/___/___ mo. day year

Signature

RRC District Office Action

▶ Expiration date ___/___/___ mo. day year   District Director   Date

Figure 7.6. Railroad Commission of Texas report form for plugging and abandonment

Describing the Reservoir

Reservoir development proceeds under the plans of the companies involved. These plans are based on the description of the reservoir provided by the geologist or the geology department. The form of description will vary according to company policies and the need for specific information by management or other departments: geophysics, engineering, geology, land and leasing, production. While the description may vary to satisfy the needs of these departments (or other clients if the geologist is working independently), it will also vary according to the type and quality of information available and the particular geologist who assembles the description.

Mapping the reservoir stratum. In practice, the map is the medium that draws the most geological evidence into recognizable form. Subsurface mapping requires painstaking effort in collecting and plotting data. The first source of possible error is at the point of collection. Each record must state the elevation, location, operator, driller, and date. Description of rocks penetrated, elevation at the top of each formation drilled, and total depth of the well are essential facts. The geologist goes to the field as often as needed to assure dependable logging of the well and analyzes the information in hand for reliability and validity. How accurate are the depths logged for the sampled cuttings? How do these depths correlate with those shown by the electric logs? When the limestone was logged, was it really dolomite? Did the core lab know the drilling direction was N 70° W and not the N 50° W listed in the report? And so on. After the data has been determined to be reasonably valid, mapping will begin.

Subsurface mapping can be thought of as similar to surface mapping. The top of the stratum to be mapped is considered the new surface. Everything above the new surface disappears. Unless the new surface is an erosional unconformity, the topography may seem a little strange. The sedimentary beds are rather flat, parallel bodies, grading gently in attitude and substance. With few visual referents, mapping calls for judgments based on slim evidence. A 12½-inch diameter core from a well on 20-acre spacing has sampled only 0.000094% of the reservoir bed. A wireline log samples scarcely more.

Types of maps. Any number of types of maps are possible. Any quality that can be isolated, quantified, and located can be mapped. Among the maps used in the petroleum industry are —

1. Structural
2. Isopach
3. Lithofacies
4. Porosity
5. Permeability
6. Paleogeographic
7. Paleogeologic
8. Paleotectonic
9. Paleoenvironment
10. Biofacies
11. Heat flow
12. Migration
13. Water composition
14. Pressure
15. Rates of production
16. Accumulated production
17. Valuation
18. Risk value
19. Development/ownership

Because there are such a number of combinations of useful, mappable characteristics and such a number of methods to derive each map, only the first three can be considered in some detail. Many excellent sources for further study are available in libraries.

Structure and isopach maps. Subsurface structure maps are drawn early in the development program and move beyond the preliminary structure map contoured before drilling the exploration well. At each well the depth from the surface to the top of the formation of interest is determined. The drilling depth is converted to a datum (fig. 7.7). Sea level is the catholic reference datum, but some other horizon is occasionally chosen. The datum-referenced values are plotted on a base map and then contoured.

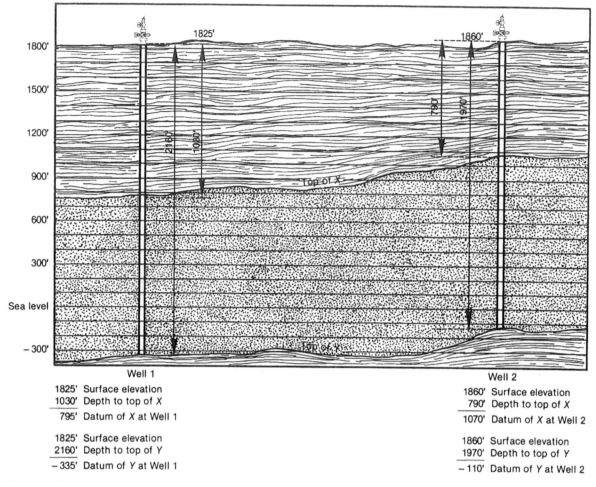

1825' Surface elevation	1860' Surface elevation
1030' Depth to top of X	790' Depth to top of X
795' Datum of X at Well 1	1070' Datum of X at Well 2
1825' Surface elevation	1860' Surface elevation
2160' Depth to top of Y	1970' Depth to top of Y
−335' Datum of Y at Well 1	−110' Datum of Y at Well 2

Figure 7.7. Method of determining datum elevation

TABLE 7.1

WELL DATA FOR FIELD BEING DEVELOPED

Well Number	Surface Elevation	Depths		Datum Elevations	
		X	Y	X	Y
1	1825	1030	2160	795	−335
2	1835	835	2105	1000	−270
3	1860	790	1970	1070	−110
4	1890	670	2020	1220	−130
5	1845	815	2285	1030	−440
6	1820	720	2250	1100	−430
7	1865	745	2215	1120	−350
8	1840	950	2640	890	−800
9	1835	840	2465	995	−630
10	1845	840	2570	1005	−725
11	1875	995	2685	880	−810
12	1860	1100	3035	760	−1175

SOURCE: L. W. LeRoy and Julian W. Low, *Graphic Problems in Petroleum Geology.* Copyright 1954 by Harper & Row, Publishers, Inc. Reprinted by permission of the publishers.

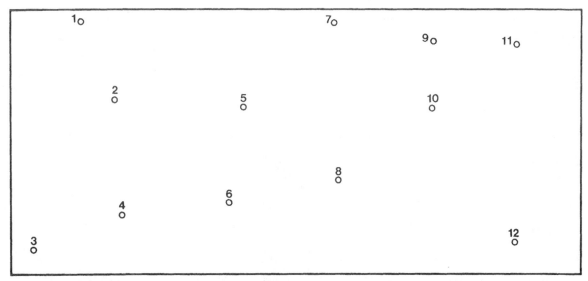

Figure 7.8. Well locations (From *Graphic Problems in Petroleum Geology* by L. W. LeRoy and Julian W. Low. Copyright 1954 by Harper & Row, Publishers, Inc. Reprinted by permission of the publishers.)

Now consider a field being developed. According to table 7.1, twelve wells have been drilled (fig 7.8). To draw the structure of the top of X, a datum (sea level) elevation must be established for X.

In contouring, each line has a constant value, and the spacing of the lines shows the rate of change in the quantity being contoured. In a structure contour map the distance between the lines indicates the slope. Closely spaced lines show a steep slope, while more widely spaced lines show a gradual change in elevation. One of the possible interpretations of structure at the top of formation X is shown in figure 7.9. Imagine that everything above X has been stripped away and you are standing on X (as if you were standing on the moon). Notice

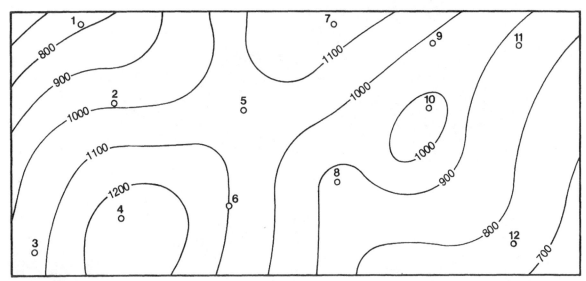

Figure 7.9. Structure on top of formation X (From *Graphic Problems in Petroleum Geology* by L. W. LeRoy and Julian W. Low. Copyright 1954 by Harper & Row, Inc. Reprinted by permission of the publishers.)

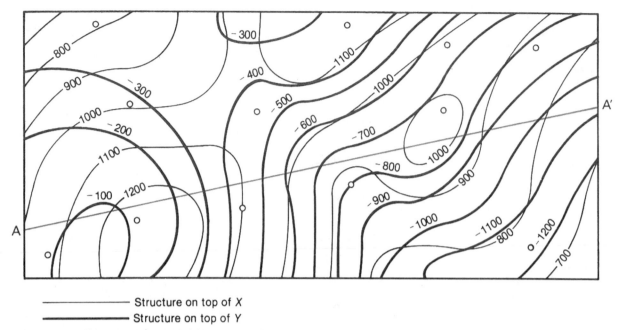

—————————— Structure on top of X
—————————— Structure on top of Y

Figure 7.10. Structure on top of formation *Y* below structure on top of *X* (Adapted from *Graphic Problems in Petroleum Geology* by L. W. LeRoy and Julian W. Low. Copyright 1954 by Harper & Row, Inc. Reprinted by permission of the publishers.)

the three hills sloping away toward the NW and SE corners. Make up your own lithology for this one.

Look below formation *X* to the top of *Y* (fig. 7.10). What does the top of *Y* look like? When a cross section is taken of the *X* and *Y*

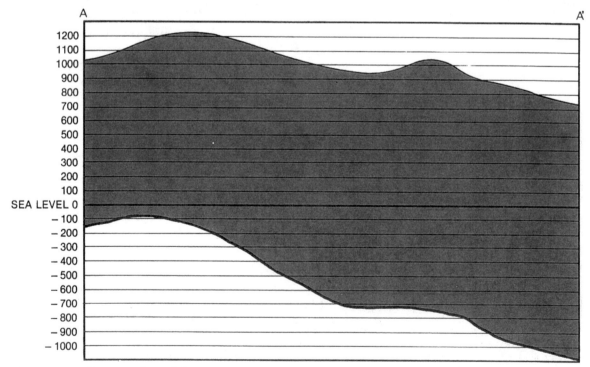

Figure 7.11. Cross section of figure 7.10

TABLE 7.2

THICKNESS OF INTERVAL X (X − Y)

Well Number	Datum Elevations		Interval between
	X	Y	X and Y
1	795	−335	1130
2	1000	−270	1270
3	1070	−110	1180
4	1220	−130	1350
5	1030	−440	1470
6	1100	−430	1530
7	1120	−350	1470
8	890	−800	1690
9	995	−630	1625
10	1005	−725	1730
11	880	−810	1690
12	760	−1175	1935

SOURCE: L. W. LeRoy and Julian W. Low, *Graphic Problems in Petroleum Geology*. Copyright 1954 by Harper & Row, Publishers, Inc. Reprinted by permission of the publishers.

values, the structure and thicknesses of *X* may be easier to see (fig. 7.11).

Go back to figure 7.10 to see the thicknesses of the interval of *X*. When the contour lines of formation *X* intersect the contour lines of formation *Y*, the elevations of both are known. For this reason, the point of each intersection is called a *control point*. At each point, subtract the datum elevation of *Y* from *X*. The difference is the thickness of *X* (table 7.2). The contoured result is shown in figure 7.12.

It helps at this point to wipe the slate clean. Leave only the isopach contours

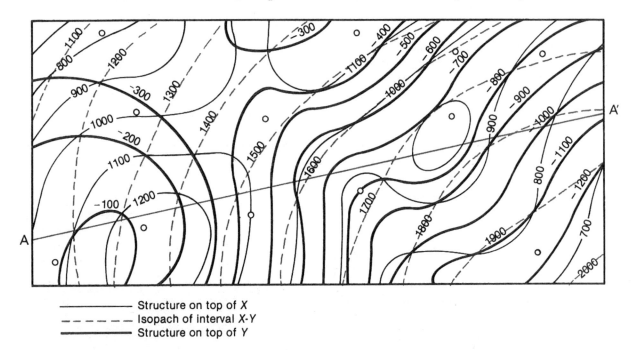

————— Structure on top of X
— — — — — Isopach of interval X-Y
————— Structure on top of Y

Figure 7.12. Isopach of interval *X-Y* on top of structure *X* and *Y* (Adapted from *Graphic Problems in Petroleum Geology* by L. W. LeRoy and Julian W. Low. Copyright 1954 by Harper & Row, Inc. Reprinted by permission of the publishers.)

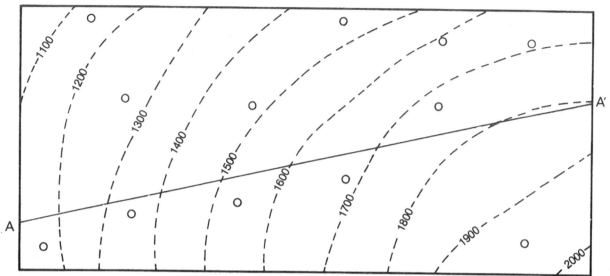

Figure 7.13. Isopach of interval *X-Y* (From *Graphic Problems in Petroleum Geology* by L. W. LeRoy and Julian W. Low. Copyright 1954 by Harper & Row, Inc. Reprinted by permission of the publishers.)

(fig. 7.13). Compare figure 7.14, a graph of the thickness along line A–A' of the isopach map, with figure 7.11, a structural cross section of the same interval.

When a well is drilled through 80 feet of sandstone, it appears that the sandstone is 80 feet thick. But if the formation is dipping away from the horizontal, the true thickness is not that great. The drilled thickness multiplied by the cosine of the dip angle will yield the true thickness (fig. 7.15), a line perpendicular to the bedding plane. Assume that the figures in table 7.1 yield true thicknesses.

Try to imagine again the surface of Y (see fig. 7.10). Stand on it and watch X being deposited. What conditions would be needed to deposit the sand X in the thickness and shape that it is? You cannot tell from the maps, because the current structure of Y as shown on the map is probably not what it was when interval X was first laid down. The strata have since undergone changes that probably included some structural deformation. The isopach can give clues, but not the whole story.

Interpreting maps and using the interpretations for decision making obviously require study of factors beyond those on the map. Among the studies shown to be most

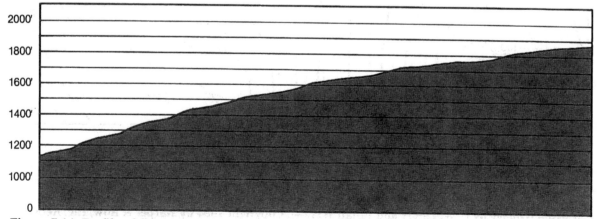

Figure 7.14. Profile of figure 7.13

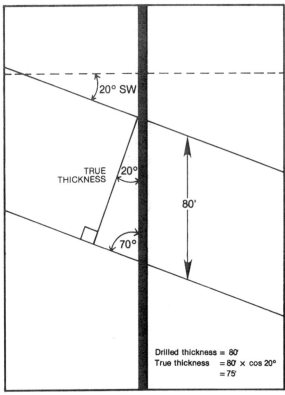

Figure 7.15. Method of determining true thickness of a formation

useful are those that cover the deposition of the reservoir stratum.

Depositional environment. The sedimentary environment concerns both erosional and depositional processes. The basic hydrodynamic, biological, and chemical conditions strongly affect both processes. Erosional processes leave such features as wave-cut sea cliffs, meandering stream channels, mesas, and buttes. Depositional processes can build features such as dunes, reefs, beaches, deltas, and bars. These features of topographic expression produced by either erosional or depositional processes can be called *geomorphic units*, or landforms.

Each geomorphic unit, beach or whatever, is characteristic in gross aspect and also possesses characteristic sedimentary structures as a result of depositional processes. Sand grain orientation, ripple marks, graded bedding, effects of animal and plant life, and precipitates can all be associated with specific environments.

Conceptual reconstruction of an ancient environment (fig. 7.16) first requires detailed

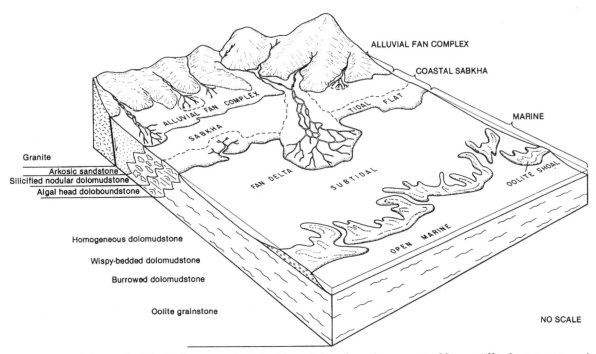

Figure 7.16. Schematic block diagram showing depositional environment of lower Ellenburger group in the Ordovician period, Puckett field, Pecos County, Texas (From *Atlas of Major Texas Oil Reservoirs* by W. E. Galloway, et al., 1983, Bureau of Economic Geology, The University of Texas at Austin)

studies of sedimentary structures. From the structures, interpretations of the hydrodynamic conditions can be made. The geometry and the vertical and lateral relationships of various depositional units can be compared with modern analogs. Reconstruction of basin configuration, paleoclimate, paleogeography, and tectonics can be undertaken, and geochemical and biological studies conducted.

Care must be taken, however, in projecting the modern environment analog to the past. A delta today certainly functions in similar fashion to past deltas, but not exactly. A sand grain of 1.0-mm diameter will be transported in water flowing at 10 cm/s yesterday as well as today. The "present is the key to the past." But each environment is also unique, and many environments of the past do not exist today. In pre-Devonian times there were no widespread land plants or animals. The erosional environment must have been considerably different. A river delta of the Ordovician period could not be just like a delta of the Cretaceous period. Tropical epeiric seas, of the type that produced the bioherms now drilled into in West Texas—the Kelly-Snyder field, for example—do not exist today.

Sequences of deposition are very important to understanding a depositional environment. Lateral relationships show a relatively thin layer of stratigraphic detail over a fairly wide geographical area. The contemporaneous facies can be seen and used for environmental reconstruction. For example, a shelf mud facies on one side of a sand body with lagoon facies on the other side indicates that the sand body might have been a barrier island. Current directions and sediment dispersal patterns may be surmised and used for visualizing basin configuration. Vertical relationships show thicknesses of the units and how they have moved with time. Together, the vertical and the horizontal patterns show the nature of transgression and regression.

The analytic process thus moves from the general regional picture to greater and greater detail, the microscopic picture, and back. Any piece that does not fit must be

carefully reexamined. When the reservoir is finally understood in its environment of deposition and structural changes over time, the search for additional pockets full of hydrocarbons becomes easier. Maps incorporating the information gained from the studies are made.

Lithofacies maps. Mapping facies changes can often clarify aspects of the reservoir that do not show in an isopach or a structure contour map. Figure 7.17 shows an isopach of a formation penetrated by twenty wells. The formation thickens to over 800 feet, but is not uniformly productive. What if each well is centered on 400

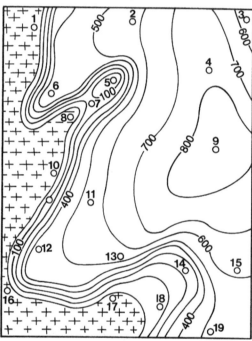

Figure 7.17. Isopach map of a formation penetrated by twenty wells, with contour interval of 100 feet (From *Graphic Problems in Petroleum Geology* by L. W. LeRoy and Julian W. Low. Copyright 1954 by Harper & Row, Publishers, Inc. Reprinted by permission of the publishers.)

acres? A good discovery well would call for development. With 40-acre spacing, up to nine additional units could be drilled around each discovery. With further analysis a more efficient development program could be undertaken.

TABLE 7.3

LITHOFACIES DATA

Well Number	Total Thickness	Aggregate Thicknesses			
		Sandstone	Shale and Silt	Carbonates	Evaporites
1	0	0	0	0	0
2	570	0	230	340	0
3	580	0	90	490	0
4	775	0	110	650	15
5	90	0	90	–	0
6	430	Trace	400	30	0
7	130	0	130	–	0
8	0	0	0	–	0
9	860	0	140	650	70
10	0	0	0	0	0
11	540	0	390	150	0
12	420	25	395	0	0
13	530	10	500	20	0
14	280	18	222	40	0
15	620	8	320	292	0
16	0	0	0	0	0
17	0	0	0	0	0
18	510	22	200	270	18
19	85	10	75	–	0
20	210	12	198	–	0

SOURCE: L. W. LeRoy and Julian W. Low, *Graphic Problems in Petroleum Geology.* Copyright 1954 by Harper & Row, Publishers, Inc. Reprinted by permission of the publishers.

Table 7.3 shows that the formation is made up of four lithologies. Notice that the thicknesses of the four component rock members add up to the total thickness. Well 18, for example, penetrates the reservoir stratum where its true thickness equals 510 feet. Of that 510 feet, 18 feet are evaporites, 270 are carbonates, 200 are shales, and 22 feet are sandstone. Each lithology need not be continuous, but may be the sum of several beds within the formation. One 22-foot layer of sandstone is the same as four 5½-foot layers. These four sand members could be identified and mapped separately. Such fine correlation work must wait until field development provides more detailed information. The data would come from electric logs run on each well, and a great deal of the geologist's time would be involved in doing the study. The very detailed correlation work would not be done unless the expected return warranted the investment.

An isopach of one of the several lithologies in a stratigraphic unit is called an *isolith*. Figure 7.18 shows an isolith of the

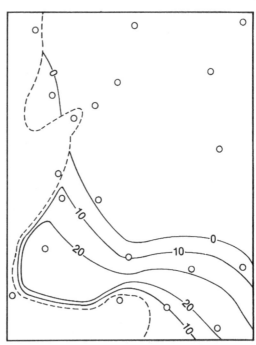

Figure 7.18. Sandstone isolith with contour interval of 10 feet (From *Graphic Problems in Petroleum Geology* by L. W. LeRoy and Julian W. Low. Copyright 1954 by Harper & Row, Inc. Reprinted by permission of the publishers.)

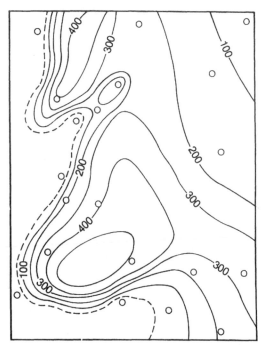

Figure 7.19. Shale and silt isolith with contour interval of 100 feet (From *Graphic Problems in Petroleum Geology* by L. W. LeRoy and Julian W. Low. Copyright 1954 by Harper & Row, Inc. Reprinted by permission of the publishers.)

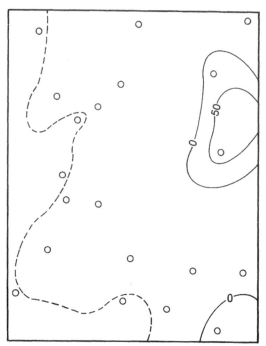

Figure 7.21. Evaporite isolith with contour of 50 feet (From *Graphic Problems in Petroleum Geology* by L. W. LeRoy and Julian W. Low. Copyright 1954 by Harper & Row, Inc. Reprinted by permission of the publishers.)

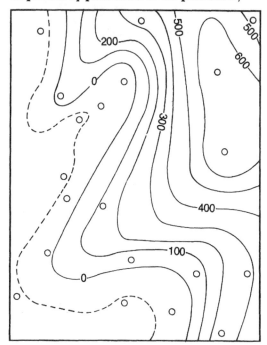

Figure 7.20. Carbonate isolith with contour interval of 100 feet (From *Graphic Problems in Petroleum Geology* by L. W. LeRoy and Julian W. Low. Copyright 1954 by Harper & Row, Inc. Reprinted by permission of the publishers.)

sandstone within the formation mapped in the isopach of figure 7.17. Check the values of the isolith lines with the sandstone data in table 7.3. Each lithology shows a distinct pattern on the isolith maps (figs. 7.18, 7.19, 7.20, 7.21). Associate the lithologies with their environments of deposition. The correlation is more clearly seen in a summary of the four isoliths, the lithofacies map shown in figure 7.22. Picture the sandstone shoreline. The clastics were bounded on the west by the margin of their source, the crystalline rocks. This supposition could be confirmed (or countered) by mineralogical study of well samples. The areas of shale indicate calm water, low-energy zones. The carbonates may provide clues to the paleoclimate and patterns of sea currents. Evaporites indicate periods of time when little fresh water reached the area.

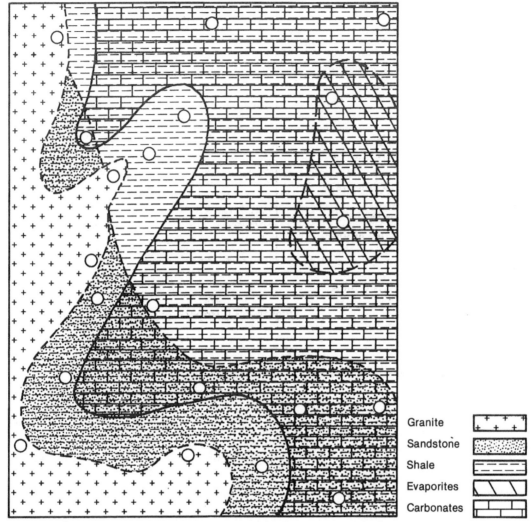

Granite

Sandstone

Shale

Evaporites

Carbonates

Figure 7.22. Lithofacies map, a summary of lithologies (From *Graphic Problems in Petroleum Geology* by L. W. LeRoy and Julian W. Low. Copyright 1954 by Harper & Row, Inc. Reprinted by permission of the publishers.)

Further studies could be undertaken. Ratio maps of carbonates to dolomites or percentage maps of the total thickness of clastics might further clarify the picture, but already the character of the formation has been unmasked. In developing this formation, the sands and carbonates will probably be drilled. Costs of development are lessened by not trying to produce from the shaly areas. Fewer dry holes will be drilled.

ESTIMATING RESERVES

As a discovery well confirms the existence of a reservoir, and as the geologist maps the field to select additional well sites, estimation of reserves becomes a concern. Accurate determination of recoverable reserves may be called for in order to

1. estimate the size of a discovery;
2. justify additional exploration work;
3. appraise property;
4. meet accounting and tax requirements;
5. settle estates;
6. appraise an exploration program; or
7. work with other producers in some type of unit operation.

Reserves cannot be known exactly, but estimates are revised as more accurate data are gathered.

Volumetric Calculation

The calculation for volume of crude oil reserves requires numbers for –

1. productive acreage, A
2. thickness of the pay zone, t
3. porosity, ϕ
4. water saturation, S_w
5. a formation volume factor, FVF
6. a recovery factor, R

The relationship of these factors in an equation for calculating volume of recoverable crude oil is

$$V = \frac{7,758 \, A \, t \, \phi \, (1-S_w) \, R}{FVF}$$

where V is the volume of recoverable stock-tank oil in barrels, and 7,758 is the conversion factor for acre-feet.

Assigning values from a hypothetical lease for each of the factors in the equation may illustrate how the calculation is done.

The *productive acreage, A,* is the area of the lease that lies within the boundaries of a field. As shown on the isopachous map (fig. 7.23), the hypothetical lease has 40 productive acres (1,320 ft × 1,320 ft = 1,742,400 sq. ft ÷ 43,560 sq. ft/acre = 40 acres).

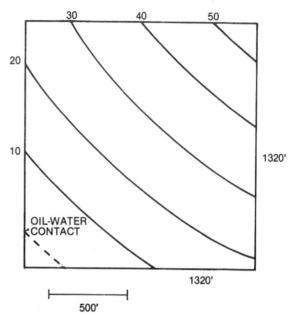

Figure 7.23. Isopach of a lease on edge of a field

Thickness of the pay zone, t, is the vertical distance within the formation that is saturated with oil and gas. Since the isopachous map (fig. 7.23) shows only the productive thickness of the formation, the average thickness can be easily calculated (50 – 0 ÷ 2 = 25 feet).

The volume (in standard U.S. units) of the lease reservoir is $A \times t = 40 \times 25 = 1,000$ acre-feet.

Porosity, ϕ, determines the amount of the total volume capable of holding fluids. Porosity is calculated from wireline log and/or core analysis. If the average reservoir porosity is calculated to be 20%, then the reservoir contains 200 acre-feet of fluids (1,000 acre-feet × 0.20 = 200 acre-feet).

Water saturation, S_w, represents the volume of pore space that is taken up by interstitial water. This water clings to the grains of the rock, making the passageways for oil smaller than the total diameter of the passageways. The saturation of water can be greater than that of oil. In the example, water saturation of the reservoir is only 35%, so 65% is saturated with fluids other than water. Only 130 acre-feet are available for oil and gas (200 acre-feet × 0.65 = 130 acre-feet).

The *formation volume factor, FVF,* is the ratio of the volume of hydrocarbons under reservoir conditions to the volume of the same hydrocarbons at standard surface conditions. It concerns the relative portions of gas, condensates, and oil and the changes that these will undergo when they are produced. For example, the petroleum might be 230°F in the reservoir and 60°F in the stock tank. The shrinkage due to this difference in temperature will be accounted for in the formation volume factor. Each type of hydrocarbon compound has its own rate of change in volume as it responds to different conditions of temperature and pressure. A sample must be taken to determine the molecular weight percent of each fraction of the petroleum. The proportions of methane, ethane, propane, butanes, pentanes, hexanes, and so on determine the response of the fluid to the specific change in temperature and pressure conditions.

The 130 acre-feet of volume calculated for the example are filled with 1,008,540 barrels of petroleum (130 × 7,758 = 1,008,540). Since the formation volume factor in this example is 1.47, then 1.47 barrels of oil in the reservoir yield only 1 barrel of oil and perhaps 800 cubic feet of gas at the surface. The resource amounts to 686,082 barrels of oil (1,008,540 ÷ 1.47 = 686,082) and 548,865,600 cubic feet of gas (686,082 × 800) dissolved under reservoir temperature and pressure conditions.

The *recovery factor, R,* concerns the volume of petroleum that can actually be produced. Much of the petroleum will cling to the pores and limit recovery to a fraction of the total. Early in the development of the reservoir, the recovery factor is often considered a function of the reservoir drive, and experience with reservoirs having analogous characteristics allows a number to be assigned. Dissolved-gas drives often recover 10% to 25% of the oil in place; gas-cap drives, up to 35%; water drives, as much as 67%. Assume that the example reservoir has an active water drive and may recover 50% of the petroleum in place.

Now the factors of the equation for calculating volume of crude oil reserves have been assigned numbers:

$$A = 40 \text{ acres}$$
$$t = 25 \text{ feet}$$
$$\phi = 0.20$$
$$S_w = 0.35$$
$$FVF = 1.47$$
$$R = 0.50$$

Substituting these values in the equation,

$$V = \frac{7{,}758 \times 40 \times 25 \times 0.20 \,(1 - 0.35)\, 0.5}{1.47}$$

$$V = 343{,}041 \text{ bbl oil.}$$

From each barrel of oil produced out of the example reservoir, 800 cubic feet of gas come out of solution by the time the gas has reached the surface. Therefore, 274,433 Mcf of gas are produced (343,041 × 800 = 274,433,000 cubic feet).

To calculate the value of reserves, a dollar figure must be assigned. The figure need not be the current price, but should represent the present value of the recoverable oil and gas over the productive life of the reservoir. At $25 per barrel of crude oil and $4 per thousand cubic feet of gas, the approximate value of the reserves under consideration would be

343,041 bbl oil @ $25.00/bbl =	$ 8,576,025.00
274,433 Mcf gas @ $4.00/Mcf =	$ 1,097,732.00
	$ 9,673,757.00

While millions of dollars always seem impressive, they may or may not add up to a profit. The gross recovery must be adjusted for royalty and overrides. If the operator has 50% of the working interest, the net recovery might amount to 142,945.18 barrels of crude and 114,356.14 Mcf of gas, or about $4,173,999.30 for a future gross income. Considering costs, future costs, taxes, and assorted headaches against the rates of production and return, it might be easier to regard this as a dry hole, ready to plug and abandon. Then again, if no special problems develop, the well might provide several people an income over the years.

Material Balance

As a field is developed, information from the production history of the wells becomes available. The three graphs in figure 7.24 show production from a reservoir under solution-gas drive. In this reservoir, as in all reservoirs, when fluid is produced at the surface, the fluids still in the reservoir adjust to the new conditions. They expand from the high-pressure state in response to the lower pressure around the well. Under the lower pressure, gas comes out of solution, increasing the ratio of gas to oil in the reservoir, decreasing drive pressure and oil mobility. Knowing the pressures and volumes of gas, oil, and water as produced under reservoir conditions allows use of the *material balance* equation. All forms of material balance are expressed by the relationship

Original stock-tank oil in place = net volumetric fluid withdrawals ÷ volumetric gas expansion of reservoir fluid per unit of stock-tank oil.

This equation takes into account such variables as reservoir fluid volumes, reservoir pressures and temperatures, compressibilities, stock-tank volumes, and water encroachment into the reservoir. Calculation of the original oil in place is one use of the equation. Other uses include determination of the amount of drive-water influx and predictions of future production behavior. Material balance calculations must be carried out at several points of time over the life of a reservoir to assure reasonable accuracy.

MODELING

Analog Models

Early attempts to predict the behavior of a reservoir were made by analogy. By construction of three-dimensional models considerable insight can be obtained and used to solve various production problems, but prediction requires keeping track of too

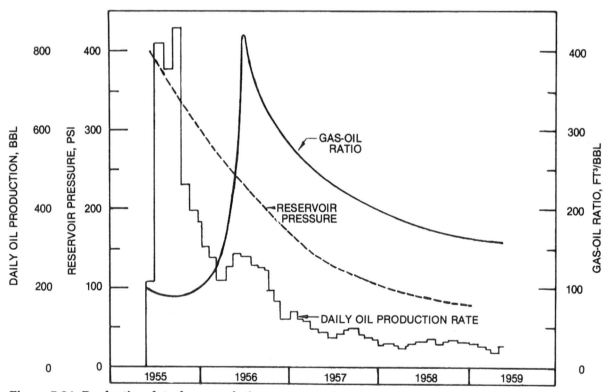

Figure 7.24. Production data from a solution-gas reservoir

TABLE 7.4

RESERVOIR MODELING PARAMETERS

Reservoir Geometry	Fluid and Well Properties	Rock Properties
Structure and isopach Quantitative log analyses Size and location of grid blocks	Production data Skin character Effective radii	Relative permeability Capillarity Saturation Oil-water contact

many variables. The geologist can only compare situations—reservoirs with similar characteristics behaving similarly. Predictions by analogy are uncertain, however, inasmuch as each reservoir is unique.

Simulation

Expressing reservoir parameters (table 7.4) in mathematical form allows precise syntheses. Reservoir modeling is a technique made practical by the computer. From appropriate geologic maps and engineering studies, data are digitized and entered into a computer program. The reservoir is divided into cells, grids, or blocks of uniform size, and the parameters are identified as discrete within each grid block. The values for each parameter are known at the present time and over the history of production to the present. Each grid is assigned its proportion of the material balance of the reservoir. The material balance equations are set up to predict future rates of production.

The program manipulating the data is called the *model*. The program expresses the parameters in mathematical form. Calibrating the model involves adjusting the program until a run matches the production history. The program is modified every 3 years or so as new pressure and production data become available. The prediction run asks the computer to explain the effect of a set of "what if's": What if we add ten wells to this section, change the production rates for these wells, and begin injecting saline water at such and such a rate in those wells?

In order to solve the problems, the program must solve each step of time within each grid first and then relate all of the grids before going to the next step. With perhaps

2,500 cells to a model, solving the finite-difference equations requires a sophisticated computer.

In final form the program will predict production over time and allow decisions to be made. With the information from figure 7.25 would you begin preparing for investment in a waterflood project?

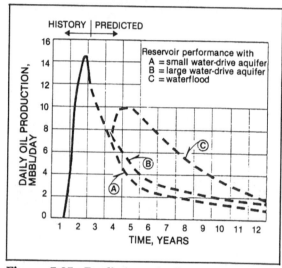

Figure 7.25. Prediction of oil producing rates (From "Reservoir Simulation Models: An Engineering Overview," by H. M. Staggs and E. F. Herbeck, *Journal of Petroleum Technology*, December 1971. Copyright 1971 by SPE-AIME.)

TABLE 7.5

IMPROVED RECOVERY METHODS

Method of Recovery		Process	Use
WATERFLOODING	Water	Water is pumped into the reservoir through injection wells to force oil toward production wells.	Method most widely used in secondary recovery.
IMMISCIBLE GAS INJECTION	Natural gas, flue gas, nitrogen	Gas is injected to maintain formation pressure, to slow the rate of decline of natural reservoir drives, and sometimes to enhance gravity drainage.	Secondary recovery.
MISCIBLE GAS INJECTION	Carbon dioxide	Under pressure, carbon dioxide becomes miscible with oil, vaporizes hydrocarbons, and enables oil to flow more freely. Often followed by injection of water.	Secondary recovery or tertiary recovery following waterflooding. Considered especially applicable to West Texas reserves because of carbon dioxide supplies located within a feasible distance.
	Hydrocarbons (propane, high-pressure methane, enriched methane)	Either naturally or under pressure, hydrocarbons are miscible with oil. May be followed by injection of gas or water.	Secondary or tertiary recovery. Supply is limited and price is high because of market demand.
	Nitrogen	Under high pressure, nitrogen can be used to displace oil miscibily.	Secondary or tertiary recovery.
CHEMICAL FLOODING	Polymer	Water thickened with polymers is used to aid waterflooding by improving fluid-flow patterns.	Used during secondary recovery to aid other processes during tertiary recovery.
	Micellar-polymer (surfactant-polymer)	A solution of detergentlike chemicals miscible with oil is injected into the reservoir. Water thickened with polymers may be used to move the solution through the reservoir.	Almost always used during tertiary recovery after secondary recovery by waterflooding.
	Alkaline (caustic)	Less expensive alkaline chemicals are injected and react with certain types of crude oil to form a chemical miscible with oil.	May be used with polymer. Has been used for tertiary recovery after secondary recovery by waterflooding or polymer flooding.

TABLE 7.5 – *Continued*

Method of Recovery		Process	Use
THERMAL RECOVERY	Steam drive	Steam is injected continuously into heavy-oil reservoirs to drive the oil toward production wells.	Primary recovery. Secondary recovery when oil is too viscous for waterflooding. Tertiary recovery after secondary recovery by waterflooding or steam soak.
	Steam soak	Steam is injected into the production well and allowed to spread during a shut-in soak period. The steam heats heavy oil in the surrounding formation and allows it to flow into the well.	Used during primary or secondary production.
	In situ combustion	Part of the oil in the reservoir is set on fire, and compressed air is injected to keep it burning. Gases and heat advance through the formation, moving the oil toward the production wells.	Used with heavy-oil reservoirs during primary recovery when oil is too viscous to flow under normal reservoir conditions. Used with thinner oils during tertiary recovery.

IMPROVING RECOVERY

The primary recovery phase of production makes use of the natural drive forces of the reservoir that urge the petroleum to surface. After the primary recovery phase in the development of a field, reservoir pressures are significantly lower than initial pressures. The fluids that are easily produced have already been produced. The next phase will be more challenging. How much oil remains? Where is it? Exactly why did it not come out? If satisfactory answers can be found, the reservoir may become a candidate for improved recovery.

The operator will consider any suggestion for additional recovery if it can be shown to be economic. The focus shifts on how to bring out this formerly unrecoverable resource. The methods used in the second (or third) approach to recovery are shown in table 7.5. Variations are many, and new methods will be conceived. The biotechnologies, for instance, may well provide for improved recovery by using microbial approaches. In any case, the method or methods chosen must conform to and account for the drive mechanism, fluid properties, and lithology or reservoir heterogeneities.

Drive Mechanism Enhancement

Recovery enhanced through improved drive mechanisms involves the injection of fluids for the purposes of restoring pressures and displacing oil. Moving volumes of water through the reservoir to production wells brings out more oil. The example in table 7.6 shows the beginning of a waterflood project – will it be worth trying for? The example shows promise.

Fluid Properties Enhancement

Projects to alter fluid properties to enhance production involve either reducing the petroleum viscosity or improving the ability of a fluid to carry or push the oil. Use of polymer floods that thicken water, injection steam, or in situ combustion that thin oil are techniques of enhancing fluid properties.

TABLE 7.6

Sample Calculation For Oil Recovery By Waterflooding

1. ESTIMATED INITIAL OIL SATURATION

   A. **Assumptions and Reservoir Data**

   The hypothetical reservoir has produced 12 percent of the original oil in place by solution-gas drive.

   Laboratory tests show the following oil reservoir volume factor (a factor that corrects for compressibility and gas dissolved in reservoir oil, calculated by dividing the volume at reservoir conditions by the volume at standard conditions in an oil stock tank on the surface):
   1.30 at original saturation pressure
   1.21 at current saturation pressure

   Connate water saturation (percentage of pore volume) equals 8 percent.

   B. **Formula for Initial Oil in Place**

   $$S_{oi} = \frac{1 - S_{wc}}{B_{oi}}$$

   where

   S_{oi} = initial saturation of oil in place
   B_{oi} = oil formation volume factor at original pressure
   S_{wc} = connate water saturation

   C. **Calculation with Substituted Reservoir Data**

   $$S_{oi} = \frac{1 - 0.08}{1.30} = \frac{0.92}{1.30} = 0.708 \text{ STBO/RVB}$$

   Initially, each barrel of reservoir pore volume held 0.708 stock tank barrels of oil (STBO).

2. PREDICTED RESIDUAL OIL AFTER WATERFLOODING (BARRELS)

   A. **Assumptions and Reservoir Data**

   Production will continue until the water cut reaches 99 percent, corresponding to an average water saturation of 70 percent of the reservoir pore volume.

   The formation volume factors are the same as in step 1.

   B. **Formula for Residual Oil per Barrel of Pore Volume in the Swept Part of the Reservoir**

   $$S_o = \frac{1 - S_w}{B_o}$$

   where

   S_o = residual oil per barrel of pore volume in the swept part of the reservoir
   B_o = oil formation volume factor at current saturation pressure
   S_w = average water saturation

   C. **Calculation with Substituted Data**

   $$S_o \frac{1 - 0.70}{1.21} = \frac{0.30}{1.21} = 0.250 \text{ STBO/RVB}$$

   At the end of waterflooding, each barrel of swept reservoir pore volume will retain 0.250 stock tank barrels of oil.

TABLE 7.6–*Continued*

3. PREDICTED RECOVERY FROM WATERFLOODING

A. Assumptions and Reservoir Data

Step 3 of the calculation employs mostly simple arithmetic and the results of steps 1 and 2.

B. Total Oil Recovery

Original oil in place per barrel of pore volume (PV)	0.708 STBO/RVB
Minus residual oil per barrel of PV after flooding	-0.250 STBO/RVB
Equals total oil recovery per barrel of PV	0.458 STBO/RVB

C. Total Recovery as a Percentage of Original Oil in Place (OOIP)

Total recovery per barrel of PV
Divided by OOIP per barrel of PV

$$100 \times \frac{0.458}{0.708} = 64.7 \text{ percent}$$

D. Percentage of Recovery by Waterflooding

Total recovery percentage	64.7 percent of OOIP
Minus primary recovery	-12.0 percent of OOIP
Equals additional recovery by waterflooding	52.7 percent of contacted OOIP

E. Correction for Heterogeneities

A correction formula (not shown here) accounts for reservoir nonuniformities by factoring in (1) the variation of permeability distribution and (2) the mobility ratio.

F. Corrected Prediction of Waterflood Recovery

The corrected estimate of recovery by waterflooding is 49 percent of the original oil in place.

4. PREDICTED RESIDUAL OIL AFTER WATERFLOODING (PERCENTAGE OF OOIP)

A. Assumptions and Reservoir Data

Step 4 uses data from step 3 and the assumption that the reservoir has produced 12 percent of the OOIP by solution-gas drive.

B. Residual Oil Available for Tertiary Recovery

OOIP	100 percent
Minus oil produced by solution-gas drive	- 12 percent
Minus oil produced by waterflooding	- 49 percent
Equals residual oil saturation	39 percent

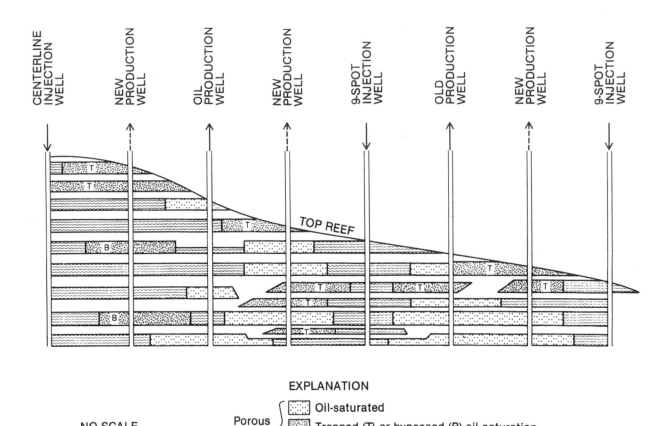

EXPLANATION

NO SCALE   Porous   ::::: Oil-saturated
               Beds   ▓▓▓▓ Trapped (T) or bypassed (B) oil saturation
                         ≡≡≡≡ Water-flushed

Figure 7.26. Schematic illustration showing the
prominent horizontal porosity stratification and
consequent zones of bypassed oil in the Canyon
reservoir, Horseshoe Atoll (From *Atlas of Major
Texas Oil Reservoirs* by W. E. Galloway, et al.,
1983, Bureau of Economic Geology, The University of Texas at Austin)

Programs Addressing
Heterogeneities

Reservoir heterogeneities can be observed
on three scales: microscopic, macroscopic,
and megascopic. Microscopic features in a
reservoir involve such factors as pore size
distribution and geometry. Microscopic
heterogeneities strongly affect the residual
water saturation and the amount of residual
oil. All improved recovery projects analyze
the microscopic features of the reservoir.

Macroscopic features include permeability
trends and stratification within the reservoir. The permeability trends are very
important for determining fluid injection

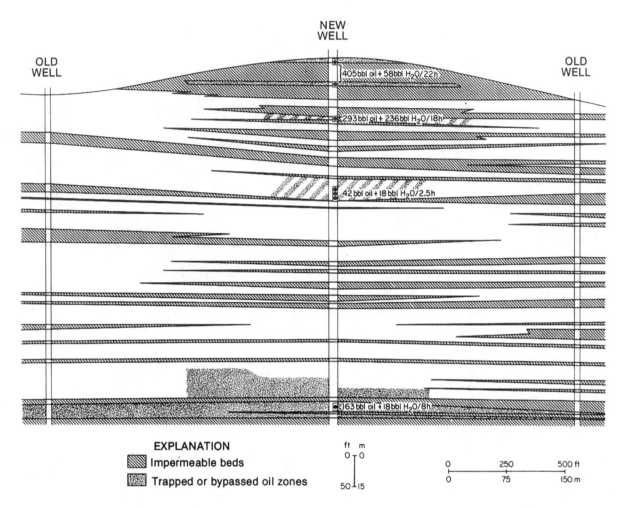

NEW
WELL

OLD
WELL

OLD
WELL

405 bbl oil + 58 bbl H₂0/22h

293 bbl oil + 236 bbl H₂0/18h

42 bbl oil + 18 bbl H₂0/2.5h

163 bbl oil + 18 bbl H₂0/8h

EXPLANATION

Impermeable beds

Trapped or bypassed oil zones

ft   m
0 ⊤ 0

50 ⊥ 15

0   250   500 ft
0   75   150 m

patterns, for modeling the reservoir, and for planning strategic infill drilling programs. Strategic infill drilling programs that spot areas not drained by the conventionally spaced drilling units can significantly improve ultimate recovery efficiencies (figs. 7.26 and 7.27).

Megascopic heterogeneities can be seen in facies changes, characteristic sealing or trapping strata patterns, and associated stratigraphic traps. The ultimate recovery efficiency of an oil play can obviously be improved by finding and identifying all of the reservoirs associated with the structure or stratum.

Figure 7.27. Production test results of an infill well drilled to recover bypassed oil in lens-shaped zones of the Canyon reservoir, Horseshoe Atoll. Systematic infill drilling has supplemented waterflood and carbon dioxide injection to optimize recovery from this giant field. (From *Atlas of Major Texas Oil Reservoirs* by W. E. Galloway, et al., 1983, Bureau of Economic Geology, The University of Texas at Austin)

Knowledge of heterogeneities of the reservoir is necessary for improved recovery. Consider the reservoir in figure 7.28. In a waterflood pilot project, injection wells *B* and *C* were drilled to displace oil to wells *A* and *D*. Well *D* performed as expected, but well *A* showed no increase in production. Going back over all of the information led to the discovery that the sands were not massive, or uniform. Part of the reservoir was

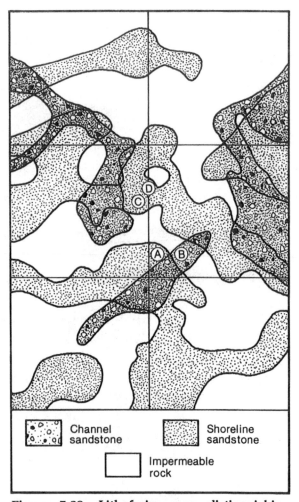

Figure 7.29. Lithofacies map distinguishing shoreline sandstone from channel sandstone

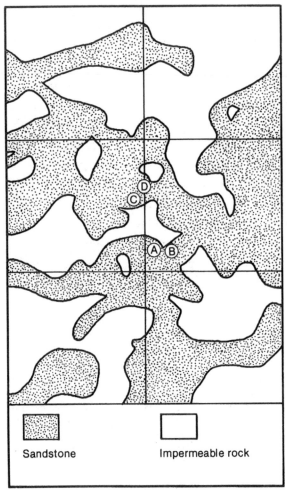

Figure 7.28. Lithofacies map of a sandstone reservoir

producing from sand deposited in the fluvial channel, and part of the production was from sand deposited on a shoreline. Permeability within each deposit was high, but between each deposit permeability was low, spelling failure for the waterflood. Remapping the isoliths (fig. 7.29) showed not only why the injection project could not work, but also the way to success.

PAST, PRESENT, AND FUTURE

The modern world runs on oil. From humble beginnings as mummy preservative, patent medicine, and lamp fuel, "rock oil" has stimulated and supported a quantum leap in man's ability to shape the environment. Plentiful in supply, easy to transport and use, the concentrated energy of hydrocarbons burns relatively clean and with high efficiency. Although wood and coal ushered in the Industrial Revolution, petroleum fueled the age of the automobile and the airplane and gave us plastic in abundance. Oil and gas now provide about two-thirds of the energy used by mankind.

But the dependence on oil and gas has placed society in an uncomfortable situation. On the one hand, it would be inconceivable to go back to the levels of energy use of the days when petroleum did not fuel civilization. Each person routinely uses more energy, and the population continues to grow. On the other hand, the resource is finite. The oil and gas will run out. Strides in efficiency may delay the time, but to continue to grow as an energy-using species, man must make the transition to other energy resources.

At some time in the future people will have to live without the energy of petroleum. The opportunity exists to use fossil fuels as a bridge to a future dependence on renewable energy sources, but the bridge will not build itself. Those who search for oil or gas will need to understand geology, the science of the earth that leads to understanding how and where oil and gas accumulate. As the search continues to grow more difficult, the role of the geologist will become even more important.

HISTORY

People have used oil and gas far longer than they have used petroleum geology to find it. Some 5,000 years ago, the temples of Ur were caulked with petroleum pitch, probably following construction practices already rich with tradition. Hammurabi's trade commission regulated the sales prices of petroleum products in Babylon. When Zarathustra visited the "eternal fires of Persia" at Baku in the sixth century B.C., 23 centuries before its first refinery, the springs had probably endured visitors carrying away little flasks full of the oil for thousands of years. Not until 1723 would Baku have a refinery. Around 200 B.C. in Szechwan, China, gas was delivered by bamboo pipelines for fires to evaporate brine. King Edward III of England (1327–77) recorded the first use of the term *petroleum* when he bought 8 pounds of it. Sir Walter Raleigh found asphalt in Trinidad in 1595.

The Seneca Indians were trading oil to the colonists in New York by 1768. The uses of oil and gas for heat and light, as construction material, and in medications, balms, and preservatives, were well established long before geologists began to argue over petroleum occurrence and means of discovery.

Early Development of the Industry

Until the middle of the nineteenth century, demand for petroleum was satisfied by production from seeps and pits. Although brine wells near Parkersburg, West Virginia, produced oil as early as 1820, no one could see the value of the crude. In 1829, a well drilled for brine near Burkesville, Kentucky, struck oil at 175 feet and flowed for 3 weeks. The oil ran down a creek to the Cumberland River, burning all the way for 40 miles. Tagged by locals as the "Great American Well," it produced "American Oil," which was bottled for medicinal purposes until 1860.

The petroleum industry began in response to the demand for fuel to illuminate the night. An oil crisis of the period had driven prices to $20 per barrel—whales not getting any easier to catch—and a new product, kerosine, stepped in to meet the demand. Distilled not from oil but from coal, kerosine was found to work well in lamps. Gas from coal was even more popular than the coal oil and was used for community street lighting. Because of the demand, it did not take long for distillers and marketers to consider petroleum as an alternate source material for both kerosine and gas.

The well credited with beginning the petroleum industry was drilled on the sound advice of a chemist. Professor Silliman of Yale advised a group of investors in 1855 that petroleum might be refined into kerosine. (Ten years later he would daringly predict that "California will be found to have more oil in its soil than all the whales in the Pacific Ocean.") One "Colonel" Drake was hired to direct the new Pennsylvania Rock Oil Company. He employed a black-smith with some salt-well experience, and they drilled on a site near a seep at Titusville, Pennsylvania. The driller, "Uncle" Billy Smith, brought in the Drake well at 69½ feet on Sunday, August 28, 1859. Once on the pump, the well produced 35 barrels of oil per day.

Supply to the burgeoning kerosine industry was assured by drilling wells. By 1861 so many wells had been drilled that the price of crude dropped to 10¢ per barrel. Even at that, wells continued to be drilled because kerosine was useful, convenient, and affordable. In 1870 John D. Rockefeller incorporated the Standard Oil Company of Ohio to refine and market products, and the petroleum age was ready for liftoff.

Oil companies drilled near seeps, followed creeks, indulged diviners of all types—and supplied 90% of the world's petroleum by 1870. "Creekology" enjoyed a measure of success due to the trellis drainage pattern of the Appalachian region. Since the major stream valleys there follow the strike of the beds, some productive anticlines were found by drilling near the creeks. Wherever oil was discovered, drillers surrounded the discovery with other wells, trying to capture their share before it was gone.

Oil was seen as transient in character, flowing through subterranean streams and lakes, here one night and gone the next. Defined as a mineral by legal fiction, oil was deemed private property. In 1889, a Pennsylvania court enunciated the rule of capture with the decision that "oil, and still more strongly gas, may be classed by themselves . . . as minerals *ferae naturae* . . . . When they escape, and go into other land, or come under another's control, the title of the former owner is gone."

Private ownership and the rule of capture have been outstanding stimuli to exploration. In spite of all the hopes and dreams dashed by drilling at ridiculous locations, the nearly random pattern of holes punched across the United States has led to the discovery of more reservoirs here than in any other country.

Development of Petroleum Geology

Serious geology of the nineteenth century stayed mostly in the realms of academia and state surveys. Petroleum did not draw much attention from geologists until after it became valuable. The fact that some oil and gas had been produced from wells was not enough to justify the building of an entire industry. Supply had to be assured by more than luck. A few geologists took the opportunity to apply science to the problem, starting with theories of how to find oil.

In 1861, T. S. Hunt of Canada theorized that oil could be found structurally contained by anticlines. The theory was not well accepted by the practical men of the oil patch, because many of the early fields of Pennsylvania were not producing from anticlinal axes. When Professor I. C. White, applying the anticline theory, discovered oil in West Virginia in the 1880s, the idea did receive critical review. However, it was not widely adopted for another 20 years.

A geological survey began to inventory coal, iron, and limestone resources in Pennsylvania in 1874. The geologist assigned to report on the oil regions, John Carll, complained about the poor logging of wells but by 1880 had produced subsurface maps correlated by paleontological study and cross sections of oil districts in Pennsylvania. He clearly showed that reading topographic features to find productive sandstone reservoirs was a poor method of exploration compared to subsurface mapping. Carll's approach was decidedly stratigraphic rather than structural, but the use of stratigraphic methods would not return to fashion for 50 or 60 years.

The first USGS subsurface structure contour map, that of the Trenton Limestone formation of Ohio and Indiana, was produced by Edward Orten in 1889. Orten reviewed the origin and accumulation of oil and gas in the Lima field, the world's first major field to produce from carbonate rock. He was the first to use the term *trap* to describe the role of structure in the accumulation of petroleum. Orten's report was an admirable model for future studies in petroleum geology, but it would be several years before the example was followed.

The first permanent geology position with an oil company was filled by Hjalmar Sjögren in 1885. He worked for the Nobel brothers, who had substantial interest in the fields at Baku. The Nobels developed the early Russian oil industry, building refineries, pipelines, and oil tankers to make use of the new production. Sjögren reported on the structure, stratigraphy, and tectonics of the Caucasus region between the Black and the Caspian seas and worked as a consultant in the region for many years.

After competing with the hunch for half a century, geology was prepared to take a major role in development of the petroleum industry. By the start of the twentieth century the organic origins of oil had been theorized, source beds identified, migration paths discussed, stratigraphic and structural traps mapped, production problems dealt with, and professional consultation performed. Not only was geology taking a professional interest in oil, but the petroleum industry was almost ready for geology.

Development of the Profession

A 1902 USGS report based in part on previous studies by several geologists described shale source beds for oil trapped in anticlinal structures in the lower San Joaquin Valley and in the Los Angeles, Ventura, and Santa Maria basins. Unlike events in the eastern United States, here petroleum geologists developed the leads and produced the oil.

William Orcutt, hired by the Union Oil Company of California in 1898 to secure information to develop oil properties, decided to apply geology. He mapped valuable acreage in the Coalinga, Kern River, and Santa Maria fields, enabling Union's production to soar tenfold within 4 years. Orcutt became a director of the company in 1908. His example was persuasive to others in the region.

The Southern Pacific Company, a railroad company with large land holdings, considered using oil for powering its locomotives in 1892 but found the supply too doubtful. The company hired E. T. Dumble as a consulting geologist to develop its natural resources and supply fuel for the railroad. Dumble soon made the career of petroleum geology possible for many young college graduates. As vice-president of the Kern Trading and Oil Company in California, the Rio Bravo Oil Company in Texas and Louisiana, and the East Coast Oil Company in Mexico, Dumble was the largest employer of petroleum geologists in the early twentieth century. In Southern California, perhaps fifty geologists were employed by various companies by 1910. These geologists developed professional standards for exploratory reconnaissance, subsurface studies, evaluation of untested prospects, and reservoir development. In 1912, no petroleum company outside California had a geology department; by 1920, most did.

Dr. Edward Bloesch of the Union des Pétroles, which became in 1913 the first company to establish a geology department in the Midcontinent, wrote: "When I came to Oklahoma I assumed that geology was an integral part of oil operations like it was in other countries. Gradually I found out that, with rare exceptions, the oil men were ignorant about the value of geology, considered it as a fad, and made jokes about it. Therefore, we geologists had to educate the oil men and the general public on the value of geology."

Professional practice was advanced by the establishment of the Southwestern Association of Petroleum Geologists at Tulsa, Oklahoma, in 1917. The next year the group reorganized as the American Association of Petroleum Geologists (AAPG).

Everette De Golyer and Wallace Everette Pratt were two giants of geology whose careers helped shape the petroleum industry. In 1910 near Tampico, Mexico, De Golyer brought in Potrero de Llano No. 4—at 110,000 barrels per day, one of the most productive wells ever drilled. This discovery led to development of the "Golden Lane" of oil fields. De Golyer went on to head the Amerada Petroleum Company, the Geophysical Research Corporation, and Geophysical Service, Inc. De Golyer and

MacNaughton has been the best-known firm of consulting geologists for most of this century.

Pratt had a little more trouble getting started. In 1916, fresh out of college, he got a job with the Texas Company in Wichita Falls, Texas, but the reception was not especially warm. "I stood in front of his desk. He did not rise. 'Mr. Pratt, you are our new geologist, Mr. Woodruff writes me. I wonder if you realize that we don't think much of geologists around here. Your office is at the end of the hall—the last door. When I want to see you, I will send for you.'" Pratt went on to become head geologist of Humble Oil and Refining Company (Exxon) and to serve on the board of directors.

The Empire Gas and Fuel Company greatly expanded the use of geology in the Midcontinent region, even adding a subsurface department in 1918. At that time, so many men were leaving to go to war that four women were added to the staff, perhaps the first to hold positions as professional geologists.

The period from 1910 to 1930 saw the widespread use of surface mapping and a growing use of subsurface mapping for structural traps. Dr. J. A. Udden of the Texas Bureau of Economic Geology called for the systematic collection and examination of well cuttings. He stressed the importance of micropaleontology, using the "bugs" to identify stratigraphic relationships. At the risk of being derisively called zoologists by the oil men, a growing number of geologists applied Dr. Udden's methods. The quality of information improved, and the quantity increased at tremendous rates.

The utility of mapping was widely accepted by oil companies once anticlines became the chief targets. The technique of mapping structure from surface geology revealed anticlines that were otherwise barely discernible. Exploration crews went all over the world, surveying with the plane table and alidade, taking samples, and drawing maps. By World War II, at least fifty significant oil fields had been discovered in the United States by mapping structures with surface expression.

Extending the search for structure below that which is revealed on the surface was very difficult but was advocated by forward thinkers. Sidney Powers, working in the 1920s for the Texas Company in the Midcontinent area, was such an advocate of subsurface mapping that associates joked that he would "rather find a buried hill than an oil field."

The information from wireline logging allowed increased sophistication of subsurface mapping. In 1912, Professor Conrad Schlumberger, of France's Ecole de Mines, began studying electrical measurements in the earth. In 1927, he ran a resistivity log in a well in Pechelbronn, Alsace. Only 7 years later, thirty-two electrical logging crews were at work—four in Europe, five in Venezuela, four in the U.S., one in the Far East, and eighteen in the USSR. Radioactivity surveys were conducted in both cased and open holes in Texas in 1938. Neutron logging began in 1941. Wireline logging was standard practice by the end of the 1930s.

The introduction of geophysical methods and wireline logging helped in the search below the surface. The Nash dome in Texas was found by torsion balance gravity survey in 1924 and proved by the bit in 1926. Also in 1924 the first seismic survey discovery took place at the Orchard salt dome on the Texas Gulf Coast. Within 20 years, geophysical methods were used to find seventy-six major fields in the United States alone.

Subsurface exploration received a big boost when papers presented at the 1929 AAPG meeting suggested that production from some of the new West Texas wells came from reefs. Because of the complex nature of reef structures, more subsurface data were needed to confirm the idea. But searching for reefs opened new regions to exploration.

Of course, new regions were being opened without the use of geology. The East Texas field, largest ever found in the lower 48, was discovered in 1929 by a veterinarian who drew lines from all the major fields of the day and found that they intersected in East Texas. "Dad" Joiner beat the geologists to the punch by drilling the discovery well,

but Humble, Gulf, and Ohio Oil had already leased much of the area before he began drilling.

A. I. Levorsen coined the term *stratigraphic trap* after further study of unconformities in the Midcontinent region. Geologists could no longer search exclusively for the high point of a structure. Production might be found in sands flanking an anticline rather than at the top. Variations in porosity as well as variations in the structure of a formation were found to trap oil and gas. The only way to find stratigraphic traps was to study and map the stratigraphy.

Mapping below unconformities is more difficult than surface mapping, but the quantity of information in some regions made it possible. A kind of layer-cake perspective developed, with each layer above and below an unconformity having its own geology. To continue to find reservoirs, it became necessary to gain a regional knowledge of all of the layers down to the basement.

The success of the new methods of exploration and production carried the petroleum industry through the depression and made available the tremendous amount of petroleum that was to prove vital in the 1940s. While the Axis powers produced less than 500 million barrels (not including the oils produced from coal) during the war, the Allies were able to draw upon reserves of some 12 billion barrels.

Geologists were in great demand after the war, not only for exploring new regions, but also for developing older areas that continued to produce. During 1959, the one-hundredth anniversary of the beginning of the oil industry, 215 shallow gas wells, 10 oilwells, 54 shallow dry holes, 89 deep gas wells, and 33 deep dry holes were drilled in Pennsylvania, plus many wells associated with secondary recovery programs and gas storage projects.

Every aspect of the industry—exploration, drilling, production, transportation, refining, and marketing—thrived on markets that expanded steadily into the 1970s. Petroleum products became available in every nation. Production from twenty-six countries in 1947 expanded to seventy-one countries in 1979 and reached the astounding level of 22.9 billion barrels. But there are signs that these heights of consumption will not be sustained. Production had already fallen to 19.3 billion barrels by 1983. Just as courageous thinking opened new areas for production in the past, new ideas will be needed to make the discoveries of the future.

THE PRODUCTION CYCLE

Past and projected hydrocarbon production forms a curve that starts near zero in 1859, rises to a peak, and drops to near zero again when the resource is exhausted at some time in the future (fig. 8.1). The areas enclosed by each of the three lines in the graph are equal, because each represents the same oil, the total amount that could ever be produced. Neither the total amount of petroleum nor the interval of time is known. But if 1880 were marked near the

beginning of the graph at the actual production of about 30 million barrels, and if 1980 were near the top of the rate of production curve at 21.765 billion barrels, and if the rate of production were to fall at a little under 7% per year, then the production in 2080 would nearly match that of two centuries earlier. Thus, the petroleum production cycle concerns the relationships over time between the resource and its consumption.

Petroleum production worldwide increased an average of 7% per year from 1890 to 1970, doubling every 10 years (fig. 8.2). Each decade has required as much as the total amount from all previous decades. Clearly, there is a limit to the number of times this doubling can occur. Perhaps the leveling of the production curve in 1980 indicates that the limit has already been reached. Even the current rate of production cannot be sustained, for the resource is finite.

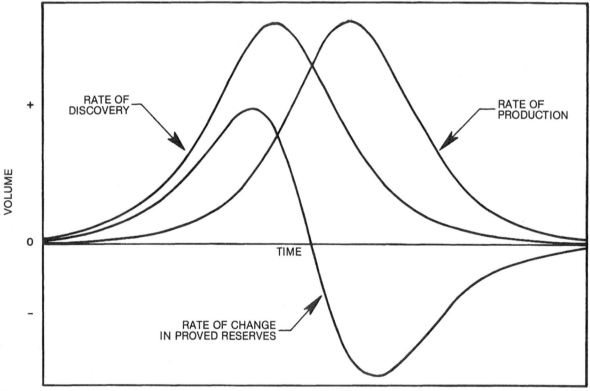

Figure 8.1. Variations of rates of discovery, of production, and of change in proved reserves of crude oil and natural gas during a complete production cycle (From M. K. Hubbert in *U. S. Energy Resources, A Review as of 1972*)

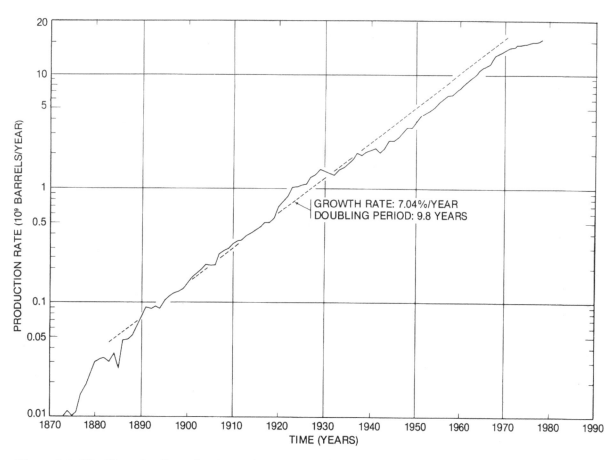

Figure 8.2. World crude oil production (From M. K. Hubbert in *U. S. Energy Resources, a Review as of 1972*)

Consumption

Over the long run, the rate of consumption and the rate of production are interdependent. An increase in consumption, or demand, brings an increase in price, leading to greater production, or supply. Conversely, increased supply leads to decreased price and greater demand. One limit to demand is the relative price and availability of substitutes. If the situation calls for a liquid fuel, then an alcohol might be substituted for oil. If the situation calls for a gas, then hydrogen might substitute for methane. The substitutes provide a price barrier that limits demand. That is, if coal can be substituted at a price (per unit of energy) lower than oil in a given situation, then demand for oil will drop. The limit to the consumption of oil will be determined by the price of coal.

One advantage of oil over the substitutes is the network of production and distribution facilities, gas stations, and roads that sustain both demand and availability. Such an infrastructure, for historical reasons, has not developed for alcohol fuels and has atrophied for coal. The ability of society to incorporate petroleum products in so many ways ensures their preeminence. This is the age of petroleum, for oil and gas are the fuels of choice.

Reserves and Resources

A natural *resource* is the total amount of an economic good in the earth, or maybe even, at some time, in space. Fossil fuel resources, all nonrenewable, include petroleum, the many types of coal, and some types of peat. Petroleum resources include oil, gas, natural gas liquids, and what are called oil shales and tar sands. Oil shales and tar sands not currently being produced are called unconventional petroleum resources. Oil or gas provided by technologies such as coal and peat gasification, biomass conversion, or other syntheses may or may not be included in the definition of unconventional petroleum resources. Organic fuels include all of the fossil fuels plus the renewable resources, primarily wood and agricultural by-products. Inorganic fuel resources involve solar, nuclear, wind, water, tidal, and geothermal energies.

The United States Geological Survey and the Bureau of Mines classify resources by the categories shown in figure 8.3. Most petroleum resources are not worth the cost of retrieval. They are in the *subeconomic* category. To produce them, more energy would be expended than would be retrieved. Furthermore, some of the resources will never be discovered. The portions of the total identified as worth producing, before being produced, are the *reserves*. The impor-

tance of reserves is that they are the source of oil production over the next several years. *Measured* reserves are those estimated from geologic evidence supported directly by engineering measurements. *Indicated* reserves are within producing reservoirs but are expected to respond only to improved recovery techniques. *Inferred* reserves are those added through revision of the calculations of reserves, extension of boundaries of reservoirs brought to light by more drilling, and occurrence of new pay zones associated with known fields. New discoveries move resources from the hypothetical and speculative categories into the reserve categories. Marginal resources are not presently recoverable because of technologic or economic factors but may become recoverable in the future. Thus change in economic climate or in technology will change the volume of reserves. However, the flow of oil through the classification system from subeconomic to

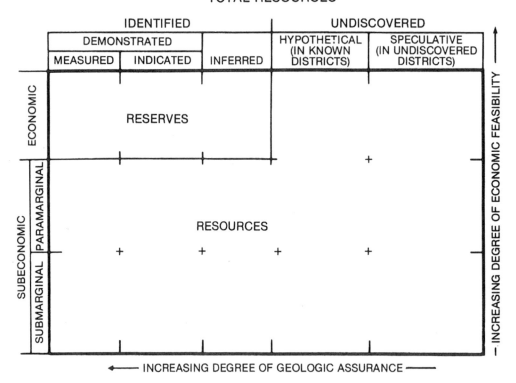

Figure 8.3. Classification of mineral reserves and resources approved by U. S. Geological Survey and U. S. Bureau of Mines (From V. E. McKelvey in *Methods of Estimating the Volume of Undiscovered Oil and Gas Resources,* Tulsa: American Association of Petroleum Geologists, 1975)

economic or from speculative to measured is not inevitable nor will it continue indefinitely.

The petroleum industry commonly uses the terms *proved, probable,* and *possible* to describe reserves, but they do not necessarily correspond to the USGS terms *measured, indicated,* and *inferred. Estimated* reserves may include proved, probable, and possible reserves along with anticipated discoveries. Each company, each country, has its own definitions and practices. Proved reserves, though, are generally accepted to be those producible at a profit using current tech-

nologies and following present economic trends. Probable and possible reserves are more doubtfully classified because of uncertainty in geologic or economic factors. Once produced, oil and gas are taken out of the reserves categories and listed as stock or inventory until they are sold and consumed.

Adjusting the numbers that describe reserves goes on constantly, since each company and each producing nation is subject to varying market forces. Because reserves are a volatile quantity, it is not enough to know that the proven reserves amount to 670 billion barrels of oil (table 8.1), and 3.2

TABLE 8.1

ESTIMATED PROVED WORLD RESERVES OF CRUDE OIL
ANNUALLY AS OF JANUARY 1
(Thousands of barrels)

Year	United States	Canada	Latin America	Middle East	Africa	Asia	Western Europe	Communist Nations	Total World
1948	21,487,685	125,000	10,085,000	28,550,000	100,000	1,300,000	50,000	6,500,000	68,197,685
1949	23,280,444	500,000	10,880,000	32,621,000	122,000	1,372,000	179,000	4,645,000	73,599,444
1950	24,649,489	1,200,000	11,400,800	32,413,000	202,500	1,557,000	288,950	4,741,000	76,452,739
1951	25,268,398	1,202,607	11,525,000	41,567,000	188,000	1,765,000	445,400	7,965,000	89,926,405
1952	27,468,031	1,376,600	12,710,850	51,320,000	175,000	1,965,500	501,500	7,927,000	103,444,481
1953	27,960,554	1,679,509	11,883,500	64,825,000	163,000	2,062,000	472,700	9,408,000	118,454,263
1954	28,944,828	1,845,422	13,182,000	78,160,000	158,000	2,583,000	641,500	9,455,000	134,969,750
1955	29,560,746	2,207,614	14,356,000	97,459,000	112,000	2,708,000	1,049,200	10,047,000	157,499,560
1956	30,012,170	2,509,534	15,515,000	126,271,000	169,000	3,000,000	1,261,000	10,833,000	189,570,704
1957	30,434,649	2,849,370	17,474,000	144,470,000	285,000	6,140,000	1,305,000	25,000,000	227,958,019
1958	30,300,405	2,874,454	21,003,500	169,566,000	814,000	8,578,000	1,304,000	26,200,000	260,640,359
1959	30,535,917	3,165,904	21,843,000	173,951,000	4,118,500	9,646,500	1,437,000	27,705,000	272,402,821
1960	31,719,347	3,497,124	24,435,600	181,436,000	7,273,500	10,175,500	1,496,000	30,002,000	290,035,071
1961	31,613,211	3,678,542	25,061,800	183,160,000	8,099,500	10,906,500	1,722,000	33,502,000	297,743,553
1962	31,758,505	4,173,569	24,706,000	188,204,000	9,709,500	10,867,600	1,735,000	34,252,000	305,407,174
1963	31,389,223	4,480,702	24,225,500	193,975,000	12,345,500	11,331,100	1,791,000	29,976,000	309,514,025
1964	30,969,990	4,881,492	24,305,000	207,368,000	16,375,500	11,621,200	1,925,600	29,500,000	326,946,782
1965	30,990,510	6,177,546	25,525,000	212,180,000	19,395,600	11,613,700	2,035,600	30,750,000	338,667,956
1966	31,352,391	6,711,237	25,170,500	215,360,000	23,049,000	11,008,750	2,085,000	33,377,000	348,118,878
1967	31,452,127	7,791,751	27,100,300	235,614,600	32,355,500	11,808,472	1,924,500	33,838,000	381,885,250
1968	31,376,670	8,168,924	26,907,250	249,209,000	42,285,250	11,815,945	2,017,000	35,773,000	407,553,039
1969	30,707,117	8,381,613	28,782,775	270,760,000	44,568,800	13,720,200	1,937,000	55,877,000	454,734,505
1970	29,631,862	8,619,805	29,179,750	333,506,000	54,679,500	13,138,150	1,779,000	60,000,000	530,534,067
1971	39,001,335[1]	8,558,980	26,185,250	344,574,900	74,757,520	14,408,648	3,708,500	100,000,000	611,195,133
1972	38,062,957[1]	8,333,087	31,558,750	367,386,000	58,886,200	15,604,500	14,222,000	98,500,000	632,553,494
1973	36,339,408[1]	8,020,141	32,601,750	355,852,000	106,402,000	14,922,260	12,082,000	98,000,000	664,219,559
1974	35,299,839[1]	7,674,150	31,640,250	350,162,500	67,303,750	15,635,040	15,990,950	103,000,000	626,706,479
1975	34,249,956[1]	7,171,229	40,578,000	403,858,200	68,299,450	21,047,700	25,814,000	111,400,000	712,418,535
1976	32,682,127[1]	6,653,002	35,368,000	368,410,570	65,085,220	21,234,230	25,487,700	103,000,000	657,920,849
1977	30,942,166	6,247,082	29,608,950	367,681,220	60,570,200	19,391,140	24,538,810	101,100,000	640,089,568
1978	29,486,402	5,970,872	40,270,000	366,166,000	59,200,150	19,749,270	26,862,500	98,000,000	645,805,194
1979	27,803,760	6,860,000	41,246,500	369,996,000	57,892,125	20,007,200	23,966,000	94,000,000	641,771,585
1980	27,051,289	6,800,000	56,472,500	361,947,300	57,072,100	19,355,200	23,476,400	90,000,000	642,174,789
1981	29,805,000	6,400,000	69,489,837	362,071,000	55,148,375	19,630,500	23,085,000	86,300,000	651,929,712
1982	29,426,000	7,300,000	84,982,270	362,839,950	56,171,630	19,150,800	24,634,500	85,845,000	670,350,150
1983	27,858,000	7,020,000	78,482,066	369,285,893	57,821,690	19,756,077	22,923,680	85,115,000	668,262,406
1984	27,735,000	6,730,000	81,675,900	370,100,800	56,907,020	18,969,400	23,019,480	84,600,000	669,737,600

(1) Figures include 9.6 billion barrels in Prudhoe Bay, Permo-Triassic Reservoir, Alaska (discovered in 1968) not yet available for production due to lack of transportation facilities.

SOURCE: 1948-1980: United States—American Petroleum Institute, Committee on Reserves and Productive Capacity
      1981-1984: United States—U.S. Energy Information Administration
      Canada—Canadian Petroleum Association (1952-1980)
      Rest of the World—*Oil and Gas Journal,* "Worldwide Oil" Issues

NOTE: Reprinted from *Basic Petroleum Data Book* with permission from the American Petroleum Institute

TABLE 8.2

ESTIMATED PROVED WORLD RESERVES OF NATURAL GAS
ANNUALLY AS OF JANUARY 1
(Billions of cubic feet)

Year	United States	Canada	Latin America	Middle East	Africa	Asia	Western Europe	Communist Nations	Total World
1967	289,333	43,450	64,550	215,070	158,155	32,450	88,582	150,000	1,041,590
1968	292,908	45,682	67,101	220,670	167,223	40,050	133,965	215,500	1,183,099
1969	287,350	47,666	62,900	223,775	168,345	52,724	141,176	343,000	1,326,936
1970	275,109	51,951	163,150	235,275	197,143	67,500	150,800	350,000	1,490,928
1971	290,746[1]	53,376	73,100	354,262	191,516	56,330	147,731	440,000	1,607,061
1972	278,806[1]	55,462	72,700	343,930	193,018	69,800	163,250	558,000	1,734,966
1973	266,085[1]	52,936	79,218	344,150	189,015	101,236	178,400	664,400	1,875,440
1974	249,950[1]	52,457	91,321	413,325	187,720	114,200	193,797	735,400	2,038,170
1975	237,132[1]	56,708	100,214	672,670	314,974	115,880	202,826	846,000	2,546,404
1976	228,200[1]	56,975	90,487	538,648	207,152	111,560	180,875	835,000	2,248,897
1977	216,026[1]	58,282	90,325	536,460	209,077	120,010	141,905	953,000	2,325,085
1978	208,878[1]	59,472	108,480	719,660	207,504	122,725	138,190	955,000	2,519,909
1979	200,302[1]	59,000	112,950	730,660	186,290	119,850	143,260	945,000	2,497,312
1980	194,917[1]	85,500	144,500	740,330	210,350	128,815	135,376	935,000	2,574,158
1981	199,021[1]	87,300	159,811	752,415	208,470	126,290	159,315	953,900	2,646,522
1982	201,730[1]	89,900	176,323	762,490	211,667	127,616	150,650	1,194,700	2,915,076
1983[r]	201,512[1]	97,000	186,591	769,730	189,423	146,247	156,736	1,283,800	3,031,039
1984	200,247	90,500	186,396	775,047	189,644	156,235	157,328	1,446,800	3,202,197

(r) Revised

(1) Figures include 26 trillion cubic feet in Prudhoe Bay, Alaska (discovered in 1968) for which transportation facilities are not yet available.

SOURCE: 1967-1980: United States – American Gas Association, Committee on Natural Gas Reserves
1981-1984: United States – Department of Energy
Rest of the world – *Oil and Gas Journal*, "Worldwide Report" Issues

NOTE: Reprinted from *Basic Petroleum Data Book* with permission from the American Petroleum Institute

quadrillion cubic feet of gas (table 8.2). The time at which the numbers apply (1984) also needs to be known. Only if all efforts to find more petroleum stopped would these numbers indicate the total amount of petroleum.

Undiscovered recoverable resources (table 8.3) are those thought to exist and expected to be producible. The estimates are based on analogy to known geologic provinces and production. About 600 basins are known on earth; 200 have not been explored. Of the 400 that have been explored, about 160 have commercial production, with 25 basins accounting for about 80% of total production.

Although small fields have played an important role in the United States, demand is now such that only the giant fields are of significance. If a field with more than 0.5 billion barrels of oil or gas equivalent is termed a giant, then about 280 fields, out of the 30,000 known, are giant fields. These giants contain about 75% of the trillion barrels of the recoverable petroleum found so far. Figure 8.4 shows the discovery rate of giant fields. The disturbing trend after 1970 may be due to political factors, technological constraints, or economic trends. Or the falling rate of discovery may indicate that half of all the oil has been discovered.

Prospects

Perhaps 40,000 people work today as petroleum geologists. Lifetimes are spent on the study of foraminifera, thermal maturation of organics, differential migration, diagenesis, paleoecology, and on and on. Each contributes knowledge to the science of the earth and to the search for energy resources. So the explorationist today draws from an abundance of methods and techniques for discovering all kinds of things, but not the oil. There is no oil-finding gadget, only hard work in a risky business.

Only a few thousand wells will be drilled in any year. Many people for many reasons decide where the wells will be drilled.

TABLE 8.3

ULTIMATELY RECOVERABLE WORLD CRUDE OIL RESOURCES
(Billion barrels)

	Known	Additional Recovery[1]	Additional Recovery[1] and New Discoveries	Ultimately Recoverable Estimate	Cumulative Production through 1975	Remaining Resource**
North America*	179.8	43–95	100–200	280–380	122	160–260
South America	68.4	20–40	52–92	120–160	41	80–120
Western Europe	24.6	5–10	25–45	50–70	3	50–70
Eastern Europe/ Soviet Union	102.4	20–40	63–123	165–225	51	110–170
Africa	75.6	15–30	45–94	120–170	21	100–150
Middle East	509.9	250–400	350–630	860–1140	85	780–1060
Asia/Oceanic	50.8	15–25	54–104	105–155	13	90–140
Unspecified	X	50–90	X	X	X	X
Total**	1000	420–730	700–1300	1700–2300	336	1360–1960

** May not add due to rounding

[1] In known fields

* Includes Mexico

SOURCE: Nehring, *Giant Oil Fields and World Oil Resources*, Rand Corp., June 1978, p. 88.

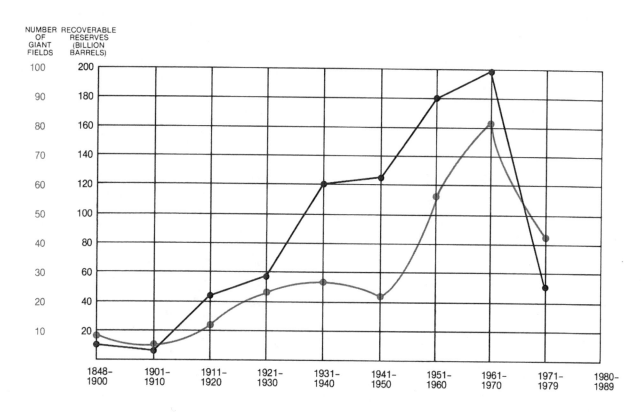

Figure 8.4. Discovery of giant oil fields (Data courtesy of Pétrole Informations)

Many wells drilled for the wrong reasons strike pay, and many more drilled for all the right reasons bottom out dry. Exploration skills are not like a recipe — "mix two pounds of core and three cans of seismic, add casing and stir for a thousand barrels." Geology does provide the framework to understand every accumulation of oil and gas. But the knowledge is only complete after the fact, if then. Whether a producer or a dry hole, every well is used by geologists to advance the science of the earth. Geology continues to slice away the misconceptions that hinder development while tempering with wisdom the misuse of the resource.

There is little doubt that the second half of the recoverable resource will be harder to find and more expensive than the first half. Three-fourths of the petroleum to be consumed in the year 2000 has yet to be discovered. Who will take the challenges of the second half, the harder half, of the petroleum age? The oil for the future will be discovered by today's geologists.

GLOSSARY

A

AAPG *abbr:* American Association of Petroleum Geologists.

absolute permeability *n:* a measure of the ability of a single fluid (such as water, gas, or oil) to flow through a rock formation when the formation is totally filled (saturated) with that fluid. Compare *effective permeability* and *relative permeability.*

accrete *v:* to enlarge by the addition of external parts or particles.

acidize *v:* to treat oil-bearing limestone or other formations with acid for the purpose of increasing production.

acoustic log *n:* a record of the measurement of porosity done by comparing depth to the time it takes for a sonic impulse to travel through a given length of formation.

acoustic survey *n:* a well-logging method in which sound impulses are generated and transmitted into the formations opposite the wellbore. Also called sonic logging.

acoustic well logging *n:* the process of recording the acoustic characteristics of subsurface formations, based on the time required for a sound wave to travel a specific distance through rock.

adhesion *n:* a force of attraction that causes molecules of one substance to cling to those of a different substance.

adsorption *n:* the adhesion of a thin film of a gas or liquid to the surface of a solid.

ad valorem tax *n:* a state or county tax based on the value of a property.

aeolian deposit *n:* a sediment deposited by wind.

aerobic *adj:* requiring free atmospheric oxygen for normal activity.

aerobic bacteria *n pl:* bacteria that require free oxygen for their life processes.

AFE *abbr:* authority for expenditure.

alidade *n:* a surveying instrument consisting of sighting device, index, and reading or recording device.

aliphatic hydrocarbons *n pl:* hydrocarbons that are characterized by having a straight chain of carbon atoms. Compare *aromatic hydrocarbons.*

aliphatic series *n:* a series of open-chained hydrocarbons. The two major classes of the series are the series with saturated bonds and the series with unsaturated bonds.

alkane *n:* also called paraffin. See *paraffin.*

allowable *n:* the amount of oil or gas legally permitted to be produced from a well per unit of time. In a state using proration, this figure is established monthly by its conservation agency. See *proration.*

alluvial fan *n:* a large, sloping sedimentary deposit at the mouth of a canyon, laid down by intermittently flowing water, especially in arid climates, and composed of gravel and sand. The deposit tends to be coarse and unworked, with angular, poorly sorted grains in thin, overlapping sheets. A line of fans may eventually coalesce into an apron that grows broader and higher as the slopes above are eroded away.

American Association of Petroleum Geologists *n:* a leading national industry organization headquartered in Tulsa, Oklahoma. Its official publication is the *AAPG Journal.*

anaerobic *adj:* active in the absence of free oxygen.

anaerobic bacteria *n pl:* bacteria that do not require free oxygen to live or are not destroyed by its absence.

analog *n:* something that is similar to something else. *adj:* representing a range of numbers by directly measurable variable quantities, such as voltages.

andesite *n:* finely crystalline, generally light-colored extrusive igneous rock composed largely of plagioclase feldspar with smaller amounts of dark-colored minerals. Compare *diorite.*

angular unconformity *n:* an unconformity in which formations above and below are not parallel. See *unconformity.*

anhydrite *n:* the common name for anhydrous calcium sulfate, CaSO₄.

anticline *n:* an arched, inverted-trough configuration of folded rock layers. Compare *syncline.*

apron *n:* 1. a body of coarse, poorly sorted sediments formed by the coalescence of alluvial or detrital fans along the flanks of a mountain range. 2. a similar body of turbidite sediments formed by the coalescence of submarine debris fans along the base of the continental slope.

aquifer *n:* 1. a permeable body of rock capable of yielding groundwater to wells and springs. 2. the part of a water-drive reservoir that contains the water.

Archie's equation *n:* the formula for evaluating the quantity of hydrocarbons in a formation.

arenite *n:* a sandstone in which less than 15% of the total volume is silt and clay.

arkose *n:* sandstone composed largely of feldspar grains and deriving from granitic source rocks.

aromatic hydrocarbons *n pl:* hydrocarbons derived from or containing a benzene ring. Many have an odor. Single-ring aromatic hydrocarbons are the benzene series (benzene, ethylbenzenes, and toluene). Aromatic hydrocarbons also include naphthalene and anthracene. Compare *aliphatic hydrocarbons.*

artificial lift *n:* any method used to raise oil to the surface through a well after reservoir pressure has declined to the point at which the well no longer produces by means of natural energy. Sucker rod pumps, gas lift, hydraulic pumps, and submersible electric pumps are the most common forms of artificial lift.

asphalt *n:* a hard brown or black material composed principally of hydrocarbons; insoluble in water but soluble in gasoline; can be obtained by heating some petroleums, coal tar, or lignite tar; used for paving and roofing and in paints.

asphaltic crude *n:* petroleum with a high proportion of naphthenic compounds, which leave relatively high proportions of asphaltic residue when refined.

asphaltic material *n:* one of a group of solid, liquid, or semisolid materials that are predominantly mixtures of heavy hydrocarbons and their nonmetallic derivatives and are obtained either from natural bituminous deposits or from the residues of petroleum refining.

associated gas *n:* natural gas that overlies and contacts crude oil in a reservoir. Also called associated free gas. See *gas cap.*

atoll *n:* a coral island consisting of a reef surrounding a lagoon.

authority for expenditure *n:* an estimate of costs prepared by a lease operator and sent to each nonoperator with a working interest for approval before work is undertaken. Normally used in connection with well drilling operations.

B

backbarrier complex *n:* the depositional environments associated with a shallow lagoon shoreward from a coastal barrier island.

backshore *n:* that part of the seashore that lies between high-tide and storm-flood level.

barefoot completion *n:* also called open-hole completion. See *open-hole completion.*

basalt *n:* an extrusive igneous rock that is dense, fine grained, and often dark gray to black in color. Compare *gabbro.*

basin *n:* 1. a local depression in the earth's crust in which sediments can accumulate to form thick sequences of sedimentary rock. 2. the area drained by a stream and its tributaries.

bed *n:* a specific layer of earth or rock, presenting a contrast to other layers of different material lying above, below, or adjacent to it.

bedding plane *n:* the surface that separates each successive layer of a stratified rock from its preceding layer where minor changes in sediments or depositional conditions can be observed.

bed load *n:* the gravel and coarse sand that are rolled and bounced along the bottom of a flowing stream. Compare *suspended load* and *dissolved load.*

bedrock *n:* solid rock exposed at the surface or just beneath the soil.

biochemical *adj:* involving chemical reactions in living organisms.

biofacies *n:* a part of a stratigraphic unit that differs in its fossil fauna and flora from the rest of the unit.

biogenic *adj:* produced by living organisms.

bioherm *n:* a reef or mound built by small organisms and their remains, such as coral, plankton, and oysters.

biomass *n:* the total mass of living organisms per unit volume per unit time.

biosphere *n:* the thin zone of air, water, and soil where all terrestrial life exists.

biotic *adj:* relating to life; biologic; relating to the actions of living organisms.

biotite *n:* a type of mica that is high in magnesium and dark in color.

blowout *n:* an uncontrolled flow of gas, oil, or other well fluids into the atmosphere.

borehole *n:* a hole made by drilling or boring; a wellbore.

bottomhole money *n:* money paid by a contributing company upon the completion of a well to a specified depth, regardless of whether the well is a producer of oil or gas or is a dry hole. The money is paid in exchange for the information received from the drilling.

bottomset bed *n:* a part of a marine delta that lies farthest from shore and consists of silt and clay extending well out from the toe of the steep delta face.

breccia *n:* a conglomerate rock composed largely of angular fragments greater than 2 mm in diameter.

brecciation *n:* the breaking of solid rock into coarse, angular fragments by faulting or crushing.

buildup test *n:* a test in which a well is shut in for a prescribed period of time and a bottomhole pressure bomb run in the well to record the pressure. From this data and from knowledge of pressures in a nearby well, the effective drainage radius or the presence of permeability barriers or other production deterrents surrounding the wellbore can be estimated.

butane *n:* a paraffin hydrocarbon, C_4H_{10}.

C

calcareous *adj:* containing or composed largely of calcium carbonate, or calcite $(CaCO_3)$.

calcite *n:* calcium carbonate $(CaCO_3)$.

calcium sulfate *n:* a chemical compound of calcium, sulfur, and oxygen. Its formula is $CaSO_4$.

caliper log *n:* a record showing variations in wellbore diameter by depth, indicating undue enlargement due to caving in, washout, or other causes.

capillarity *n:* the rise or fall of liquids in small-diameter tubes or tubelike spaces, caused by the combined action of surface tension (cohesion) and wetting (adhesion). See *capillary pressure.*

capillary pressure *n:* a pressure or adhesive force caused by the surface tension of water. This pressure causes the water to adhere more tightly to the surface of small pore spaces than to larger ones. Capillary pressure in a rock formation is comparable to the pressure of water that rises higher in a small glass capillary tube than it does in a larger tube.

capitalized *adj:* deducted from income over the years of useful life of an item purchased.

caprock *n:* 1. a disklike plate of anhydrite, gypsum, limestone, or sulfur overlying most salt domes in the Gulf Coast region. 2. impermeable rock overlying an oil or gas reservoir that tends to prevent migration of oil or gas out of the reservoir.

carbonate *n:* 1. a salt of carbonic acid. 2. a compound containing the carbonate (CO_3^{--}) radical.

carbonate mud *n:* a mud that forms on the seafloor by the accumulation of calcite particles. It may eventually become limestone.

carbonate rock *n:* a sedimentary rock composed primarily of calcium carbonate (calcite) or calcium magnesium carbonate (dolomite).

carbonation *n:* 1. a chemical reaction that produces carbonates. 2. in geology, a form of chemical weathering in which a mineral reacts with carbon dioxide (in solution as carbonic acid) to form a carbonate mineral.

carbonic *adj:* of or relating to carbon, carbonic acid, or carbon dioxide.

carbonize *v:* to convert into carbon or a carbonic residue.

cash flow *n:* the difference between inflow and outflow of funds over a period of time. Cash flow can be positive (profit) or negative (loss).

cash flow analysis *n:* an economic analysis that relates investments to subsequent revenues and also makes possible a comparison between investments. It usually includes also the general plan to be used for the figuring of federal income taxing on the investments.

catastrophism *n:* the theory that the earth's landforms assumed their present configuration in a brief episode at the beginning of geologic history, possibly in a single great catastrophic event, and have remained relatively unchanged since that time. Compare *uniformitarianism.*

cavern *n:* a natural càvity in the earth's crust that is large enough to permit human entry; commonly formed in limestone due to leaching by groundwater. Compare *vug.*

cementation *n:* the crystallization or precipitation of soluble minerals in the pore spaces between clastic particles, causing them to become consolidated into sedimentary rock.

Cenozoic era *n:* the time period from 65,000,000 years ago until the present. It is marked by rapid evolution of mammals and birds, flowering plants, grasses, and shrubs, and little change in invertebrates.

chert *n:* a rock of precipitated silica whose crystalline structure is not easily discernible and which fractures conchoidally (like glass). Flint, jasper, and chat are forms of chert.

chromatograph *n:* an analytical instrument that separates mixtures of substances into identifiable components by means of chromatography.

chromatography *n:* a method of separating a solution of closely related compounds by allowing it to seep through an adsorbent so that each compound becomes adsorbed in a separate layer.

clastic rock *n:* a sedimentary rock composed of fragments of preexisting rocks. The principal distinction among clastics is grain size. Conglomerates, sandstones, and shales are clastic rocks.

clastics *n pl:* 1. sediments formed by the breakdown of large rock masses by climatological processes, physical or chemical. 2. the rocks formed from these sediments.

clastic texture *n:* rock texture in which individual rock, mineral, or organic fragments are cemented together by an amorphous or crystalline mineral such as calcite. Compare *crystalline texture.*

clay *n:* 1. a term used for particles smaller than $\frac{1}{256}$ mm (4 microns) in size, regardless of mineral composition. 2. a group of hydrous aluminum silicate minerals (clay minerals). 3. a sediment of fine clastics.

coal *n:* a carbonaceous rocklike material that forms from the remnants of plants that were subjected to biochemical processes, intense pressure, and high temperatures; used as fuel.

cohesion *n:* the attractive force between the same kinds of molecules (i.e., the force that holds the molecules of a substance together).

compaction *n:* a decrease in the volume of a stratum due to pressure exerted by overlying strata, evaporation of water, or other causes.

compaction anticline *n:* See *draped anticline.*

complete a well *v:* to finish work on a well and bring it to productive status. See *well completion.*

condensate *n:* a light hydrocarbon liquid obtained by condensation of hydrocarbon vapors. It consists of varying proportions of butane, propane, pentane, and heavier fractions, with little or no methane or ethane.

conductor pipe *n:* a short string of large-diameter casing used to keep the wellbore open and to provide a means of conveying the upflowing drilling fluid from the wellbore to the mud pit.

conglomerate *n:* a sedimentary rock composed of pebbles of various sizes held together by a cementing material such as clay. Conglomerates are similar to sandstone but are composed mostly of grains more than 2 mm in diameter. Most conglomerates are found in discontinuous, thin, isolated layers; they are not very abundant. In common usage, the term *conglomerate* is restricted to coarse sedimentary rock with rounded grains; conglomerates made up of sharp, angular fragments are called breccia.

connate water *n:* water retained in the pore spaces, or interstices, of a formation from the time the formation was created. Compare *interstitial water.*

contact *n:* 1. any sharp or well-defined boundary between two different bodies of rock. 2. a bedding plane or unconformity that separates formations.

contact metamorphism *n:* a type of metamorphism that occurs when an intruded body of molten igneous rock changes the rocks immediately around it, primarily by heating and by chemical alteration.

continental drift *n:* according to a 1910 theory of Alfred Wegener, a German meteorologist, the migration of continents across the ocean floor like rafts drifting at sea. Compare *plate tectonics.*

continental rise *n:* the transition zone between the continental slope and the oceanic abyss.

continental shelf *n:* a zone, adjacent to a continent, that extends from the low waterline to the continental slope, the point at which the seafloor begins to slope off steeply into the oceanic abyss.

continental slope *n:* a zone of steep, variable topography forming a transition from the continental shelf edge to the ocean basin.

contour map *n:* a map constructed with continuous lines connecting points of equal value, such as elevation, formation thickness, rock porosity, and so forth.

core *n:* 1. a cylindrical sample taken from a formation for geological analysis. 2. the metallic, partly solid and partly molten interior of the earth, about 4,400 miles in diameter. *v:* to obtain a solid, cylindrical formation sample for analysis.

core analysis *n:* laboratory analysis of a core sample to determine porosity, permeability, lithology, fluid content, angle of dip, geological age, and probable productivity of the formation.

Cretaceous *adj:* of or relating to the geologic period from about 135 million to 65 million years ago at the end of the Mesozoic era, or to the rocks formed during this period, including the extensive chalk deposits for which it was named.

crooked hole *n:* a wellbore that has been unintentionally drilled in a direction other than vertical.

cross-bedding *n:* sedimentation in which laminations are transverse to the main stratification planes.

crude oil *n:* unrefined liquid petroleum.

crust *n:* the outer layer of the earth, varying in thickness from 5 to 30 miles (10 to 50 km). It is composed chiefly of oxygen, silicon, and aluminum.

crystalline texture *n:* rock texture that is the result of progressive and simultaneous interlocking growth of mineral crystals. Compare *clastic texture.*

crystallization *n:* the formation of crystals from solutions or melts.

cut fluorescence test *n:* a test involving the observation of a formation sample, immersed in a solvent, under ultraviolet light. Hydrocarbons fluoresce under ultraviolet light. If any hydrocarbons are in the sample, they will dissolve and appear as streamers or streaks of color different from the solvent.

cuttings *n-pl:* the fragments of rock dislodged by the bit and brought to the surface in the drilling mud.

cycloparaffin *n:* a saturated nonaromatic hydrocarbon compound with ring-shaped molecules, of the general chemical formula C_nH_{2n}. Also called naphthene.

D

darcy *n:* a unit of measure of permeability. A porous medium has a permeability of 1 darcy when a pressure drop of 1 atmosphere across a sample 1 cm long and 1 cm² in cross section will force a liquid of 1-cp viscosity through the sample at the rate of 1 cm³ per second. The permeability of reservoir rocks is usually so low that it is measured in millidarcy units.

database *n:* a complete collection of information, such as contained on magnetic disks or in the memory of an electronic computer.

datum *n:* a standard elevation, such as mean sea level, relative to which other elevations are measured and displayed on contour maps.

DCFR *abbr:* discounted cash flow rate of return.

decision tree *n:* a graphic representation of predicted financial gains or losses for the outcomes of several courses of action.

deformation *n:* the action of earth stresses that results in folding, faulting, shearing, or compression of rocks.

deliverability plot *n:* a graph that compares flowing bottomhole pressure of a well with production in barrels of oil per day to show the relationship between drawdown and the producing rate. The main purpose of deliverability plotting is to find the most efficient flow rate for the well.

delta *n:* See *marine delta* or *lacustrine delta.*

density log *n:* a special radioactivity log for open-hole surveying that responds to variations in the specific gravity of formations.

deplete *v:* to exhaust a supply. An oil and gas reservoir is depleted when most or all economically recoverable hydrocarbons have been produced.

depletion *n:* 1. the exhaustion of a resource. 2. a reduction in income reflecting the exhaustion of a resource. The concept of depletion recognizes that a natural resource such as oil is used up over several accounting periods and permits the value of this resource to be expensed periodically as the resource is exhausted.

depletion allowance *n:* a reduction in U.S. taxes for owners of an economic interest in minerals in place to compensate for the exhaustion of an irreplaceable capital asset.

deposition *n:* the laying down of sediments or other potential rock-forming material.

depositional environment *n:* the set of physical, chemical, and geological conditions (such as climate, streamflow, sediment source, etc.) under which a rock layer was laid down.

depreciation *n:* 1. decrease in value of an asset such as a plant or equipment due to normal wear or passing of time; real property (land) does not depreciate. 2. an annual reduction of income reflecting the loss in useful value of capitalized investments by reason of wear and tear.

development well *n:* 1. a well drilled in proven territory in a field to complete a pattern of production. 2. an exploitation well. See *exploitation well.*

Devonian *adj:* of or relating to the geologic period from about 400 million to 350 million years ago in the Paleozoic era, or to the rocks formed during this period, including those of Devonshire, England, where outcrops of such rocks were first identified.

diagenesis *n:* the chemical and physical changes that sedimentary deposits undergo (compaction, cementation, recrystallization, and sometimes replacement) during and after lithification.

diapir *n:* a dome or anticlinal fold in which a mobile plastic core has ruptured the more brittle overlying rock. Also called piercement dome.

diapirism *n:* the penetration of overlying layers by a rising column of salt or other easily deformed mineral due to differences in density.

diastrophism *n:* the process or processes of deformation of the earth's crust that produce oceans, continents, mountains, folds, and faults.

diatom *n:* any of the algae of the class Bacillariophyceae, noted for symmetrical and sculptured siliceous cell walls. After death, the cell wall persists and forms diatomite. Diatoms appeared in the Cretaceous period.

diatomite *n:* a rock, of biochemical origin, which is composed of the siliceous (glassy) shells of microscopic algae called diatoms.

differential pressure *n:* the difference between two fluid pressures; for example, the difference between the pressure in a reservoir and in a wellbore drilled in the reservoir, or between atmospheric pressure at sea level and at 10,000 feet. Also called pressure differential.

digital *adj:* pertaining to data in the form of digits; especially, electronic data stored in the form of a binary code.

diorite *n:* intrusive, or plutonic, generally coarse-grained igneous rock composed largely of plagioclase feldspar with smaller amounts of dark-colored minerals. Also known as black granite. Compare *andesite.*

dip *n:* also called formation dip. See *formation dip.*

dip log *n:* See *dipmeter survey.*

dipmeter log *n:* See *dipmeter survey.*

dipmeter survey *n:* an oilwell-surveying method that determines the direction and angle of formation dip in relation to the borehole.

dip slip *n:* upward or downward displacement of a fault plane.

directional drilling *n:* intentional deviation of a wellbore from the vertical.

disconformity *n:* an unconformity above and below which rock strata are parallel. A disconformity may or may not be parallel to these strata. See *unconformity;* compare *nonconformity.*

discounted cash flow rate of return *n:* the rate that causes the sum of the discounted outflows and inflows of funds to equal the net cash outlay in year zero of a project; used in evaluating exploration investments.

discovery well *n:* the first oil or gas well drilled in a new field that reveals the presence of a petroleum-bearing reservoir. Subsequent wells are development wells.

dissolved load *n:* in a flowing stream of water, those products of weathering that are carried along in solution. Compare *bed load* and *suspended load.*

division order *n:* a contract of sale of oil or gas to a purchaser who is directed to pay for the oil or gas products according to the proportions set out in the division order. The purchaser may require execution thereof by all owners of interest in the property.

dolomite *n:* a type of sedimentary rock similar to limestone but containing more than 50% magnesium carbonate; sometimes a reservoir rock for petroleum.

dolomitization *n:* the shrinking of the solid volume of rock as limestone turns to dolomite; the conversion of limestone to dolomite rock by replacement of a portion of the calcium carbonate with magnesium carbonate.

dolostone *n:* rock composed of dolomite.

dome *n:* a geologic structure resembling an inverted bowl; a short anticline that dips or plunges on all sides.

doodlebug *n:* (slang) a person who prospects for oil, especially by using seismology. Also called doodlebugger. *v:* (slang) to explore for oil, especially by using seismic techniques in which explosive charges are detonated in shot holes to create shock waves (taken from the resemblance of these explosions to the puffs of loose dirt thrown up by the doodlebug, or ant lion, when constructing its funnel-shaped trap).

downcutting *n:* the direct erosive action of flowing water on a streambed.

downdip *adj:* lower on the formation dip angle than a particular point.

drag fold *n:* frictional deformation of the layers above or below an overthrust fault.

draped anticline *n:* an anticline composed of sedimentary deposits atop a reef or atoll, along whose flanks greater thicknesses of sediments have been deposited and compacted than atop the reef itself. Also called compaction anticline.

drawdown *n:* 1. the difference between static and flowing bottomhole pressures. 2. the distance between the static level and the pumping level of the fluid in the annulus of a pumping well.

driller's log *n:* a record that describes each formation encountered and lists the drilling time relative to depth, usually in 5- to 10-foot (1.5 to 3 m) intervals.

drill stem test *n:* the conventional method of formation testing.

drive *n:* the energy of expanding gas, inflowing water, or other natural or artificial mechanisms that forces crude oil out of the reservoir formation and into the wellbore.

dry hole *n:* any well that does not produce oil or gas in commercial quantities.

dry hole money *n:* money paid by a contributing company on the basis of so much per foot drilled by the primary company in return for information gained from the drilling. The contribution is paid only if the well is a dry hole in all formations encountered in drilling.

dunefield *n:* an accumulation of windborne sand in that part of the seashore that lies above storm-flood level.

E

ecology *n:* science of the relationships between organisms and their environment.

effective permeability *n:* a measure of the ability of a single fluid to flow through a rock when another fluid is also present in the pore spaces. Compare *absolute permeability* and *relative permeability.*

effective porosity *n:* the percentage of the bulk volume of a rock sample that is composed of interconnected pore spaces that allow the passage of fluids through the sample. See *porosity.*

Eh *sym:* oxidation-reduction potential.

electric log *n:* also called an electric well log. See *electric well log.*

electric survey *n:* also called an electric well log. See *electric well log.*

electric well log *n:* a record of certain electrical characteristics of formations traversed by the borehole, made to identify the formations, determine the nature and amount of fluids they contain, and estimate their depth. Also called an electric log or electric survey.

emulsion *n:* a mixture in which one liquid, termed the dispersed phase, is uniformly distributed (usually as minute globules) in another liquid, called the continuous phase or dispersion medium. In an oil-water emulsion, the oil is the dispersed phase and the water the dispersion medium; in a water-oil emulsion, the reverse holds.

enhanced oil recovery *n:* 1. the introduction of an artificial drive and displacement mechanism into a reservoir to produce oil unrecoverable by primary recovery methods. EOR methods include waterflooding, chemical flooding, most types of gas injection, and thermal recovery. 2. the use of an advanced EOR method.

entrained *adj:* drawn in and transported by the flow of a fluid.

epeiric sea *n:* a shallow arm of the ocean that extends from the continental shelf deep into the interior of the continent. Also called epicontinental sea.

epoch *n:* a division of geologic time; a subdivision of a geologic period.

era *n:* one of the major divisions of geologic time.

erosion *n:* the process by which material (such as rock or soil) is worn away or removed (as by wind or water).

estuary *n:* a coastal indentation or bay into which a river empties and where fresh water mixes with seawater. Compare *marine delta.*

ethane *n:* a paraffin hydrocarbon, C_2H_6; under atmospheric conditions, a gas, one of the components of natural gas.

evaporite *n:* a sedimentary rock formed by precipitation of dissolved solids from water evaporating in enclosed basins. Examples are gypsum and salt.

expected value concept *n:* a risk analysis process that multiplies expected gain or loss of a decision by its probability of occurrence and averages all possible outcomes in order to choose the action with the highest expected benefit.

expensed *adj:* deducted from income in the year in which the expenditure is incurred.

exploitation well *n:* a well drilled to permit more effective extraction of oil from a reservoir. Sometimes called a development well. See *development well.*

exploration *n:* the search for reservoirs of oil and gas, including aerial and geophysical surveys, geological studies, core testing, and drilling of wildcats.

extrusion *n:* 1. the emission of magma (as lava) at the earth's surface. 2. the body of igneous rock produced by the process of extrusion.

extrusive *adj:* volcanic; derived from magmatic materials poured out upon the earth's surface, as distinct from intrusive rocks formed from magma that has cooled and solidified beneath the surface.

extrusive rock *n:* igneous rock formed from lava poured out on the earth's surface.

F

facies *n:* part of a bed of sedimentary rock that differs significantly from other parts of the bed.

fanglomerate *n:* coarse-grained, poorly sorted sedimentary rock derived from sediments deposited in alluvial fans; a type of conglomerate.

farmout *n:* a contract between a lessee and a third party to assign leasehold interest to the third party, conditional upon the third party's drilling a well within the expiration date of the primary term of the lease.

farm out *v:* to assign leasehold interest to a third party, with stipulated conditions. See *farmout.*

fault *n:* a break in the earth's crust along which rocks on one side have been displaced (upward, downward, or laterally) relative to those on the other side.

fault plane *n:* a surface along which faulting has occurred.

fault trap *n:* a subsurface hydrocarbon trap created by faulting, in which an impermeable rock layer has moved opposite the reservoir bed or where impermeable gouge has sealed the fault and stopped fluid migration.

faunal succession *n:* the principle that fossils in a stratigraphic sequence succeed one another in a definite, recognizable order.

feldspar *n:* a group of silicate minerals that include a wide variety of potassium, sodium, and aluminum silicates. Feldspar makes up about 60% of the outer 15 km of the earth's crust.

feldspathic *adj:* containing or largely composed of feldspar or feldspar grains.

fence diagram *n:* See *panel diagram.*

field *n:* a geographical area in which a number of oil or gas wells produce from a continuous reservoir. A field may refer to surface area only or to underground productive formations as well. A single field may have several separate reservoirs at varying depths.

fluid potential *n:* for a fluid flow, a scalar function whose gradient is the velocity of the fluid. Also called velocity potential.

fluvial deposit *n:* sediment deposited by flowing water.

flux gate *n:* a detector that produces an electrical signal whose magnitude and phase are proportional to the magnitude and direction of the external magnetic field acting along its axis; used to indicate the direction of the earth's magnetic field.

flysch *n:* a type of rock consisting of thinly bedded sandstone and shale, thought to be the result of action of turbidity currents; a succession of turbidites originating in marine depositional basins, usually near the base of the continental slope. Flysch deposits are especially common in the Alpine region of Europe.

fold *n:* a flexure of rock strata (e.g., an arch or a trough) produced by horizontal compression of the earth's crust. See *anticline* and *syncline.*

foliated metamorphic rock *n:* metamorphic rock that has a layered look not necessarily associated with the original layering in sedimentary rock.

footwall *n:* the rock surface forming the underside of a fault when the fault plane is not vertical—that is, if the dip is less than 90°. Compare *hanging wall.*

foraminifera *n:* single-celled, mostly microscopic animals with calcareous exoskeletons; mostly marine.

foreset bed *n:* a depositional layer on the steep seaward face of a marine delta that lies beyond the topset beds and is composed of finer sedimentary materials than the topset beds.

foreshore *n:* that part of the seashore that lies between low- and high-tide levels.

formation *n:* a bed or deposit composed throughout of substantially the same kind of rock; often a lithologic unit.

formation dip *n:* the angle at which a formation bed inclines away from the horizontal. *Dip* is also used to describe the orientation of a fault.

formation fracturing *n:* a method of stimulating production by opening new flow channels in the rock surrounding a production well.

formation strike *n:* the horizontal direction of a formation bed as measured at a right angle to the dip of the bed.

formation testing *n:* the gathering of pressure data and fluid samples from a formation to determine its production potential before choosing a completion method.

fossil *n:* the remains or impressions of a plant or animal of past geological ages that have been preserved in or as rock.

fossiliferous *adj:* containing fossils.

fossilize *v:* to become changed into a fossil.

fracturing *n:* shortened form of formation fracturing. See *formation fracturing.*

frost wedging *n:* the phenomenon resulting when water invades rock, then freezes, and by its expansion wedges apart the rock. Repeated freeze-thaw cycles can quickly break up any rock that has even the tiniest cracks.

G

gabbro *n:* an intrusive igneous rock with the same composition as basalt.

geochemistry *n:* study of the relative and absolute abundances of the elements of the earth and the physical and chemical processes that have produced their observed distributions.

geomorphic unit *n:* one of the features that, taken together, make up the form of the surface of the earth.

gamma-ray log *n:* a type of radioactivity well log that records natural radioactivity around the wellbore.

gas cap *n:* a free-gas phase overlying an oil zone and occurring within the same producing formation as the oil. See *reservoir* and *associated gas.*

gas chromatograph *n:* a device used to separate and identify gas compounds by their adhesion to different layers of a filtering medium such as clay or paper, sometimes indicated by color changes in the medium.

gas drive *n:* the use of the energy that arises from the expansion of compressed gas in a reservoir to move crude oil to a wellbore.

geologic time scale *n:* the long periods of time dealt with and identified by geology. Geologic time is divided into eras (usually Cenozoic, Mesozoic, Paleozoic, and Precambrian), which are subdivided into periods and epochs. When the age of a type of rock is determined, it is assigned a place in the scale and thereafter referred to as, for example, Mesozoic rock of the Triassic period.

geologist *n:* a scientist who gathers and interprets data pertaining to the rocks of the earth's crust.

geology *n:* the science that relates to the study of the structure, origin, history, and development of the earth and its inhabitants as revealed in the study of rocks, formations, and fossils.

geophone *n:* an instrument that detects vibrations passing through the earth's crust, used in conjunction with seismography. Geophones are often called jugs.

geophysical exploration *n:* measurement of the physical properties of the earth in order to locate subsurface formations that may contain commercial accumulations of oil, gas, or other minerals.

geophysicist *n:* one who studies geophysics. See *geophysics.*

geophysics *n:* the physics of the earth, including meteorology, hydrology, oceanography, seismology, volcanology, magnetism, and radioactivity.

geoscience *n:* a science dealing with the earth—geology, physical geography, geophysics, geomorphology, geochemistry.

geostatic pressure *n:* the pressure to which a formation is subjected by its overburden. Also called ground pressure, rock pressure, lithostatic pressure.

geostatic pressure gradient *n:* the change in geostatic pressure per unit of depth in the earth.

geothermal *adj:* pertaining to heat within the earth.

Gondwanaland *n:* the southern part of the supercontinent Pangaea, comprising the future land masses of South America, Africa, Antarctica, Australia, and India.

gouge *n:* finely abraded material occurring between the walls of a fault, the result of grinding movement.

graben *n:* a block of the earth's crust that has slid downward between two faults; the opposite of a horst.

graded stream *n:* a flowing stream that is stable, or in balance with its average load. It is just steep enough to carry out of its basin the amount of sediment brought in during an average-flow year.

gradualism *n:* See *uniformitarianism.*

granite *n:* an igneous rock composed primarily of feldspar, quartz, and mica. It is the most common intrusive rock—that is, it originally solidified below the surface of the earth. Its crystals are easily seen by the unaided eye.

graphite *n:* a soft black shiny mineral of pure carbon produced when hydrocarbons are subjected to high temperatures and pressures.

gravel pack *n:* a mass of very fine gravel placed around a slotted liner in a well.

gravimeter *n:* an instrument used to detect and measure minute differences in the earth's gravitational pull at different locations to obtain data about subsurface formations.

gravimetric survey *n:* the survey made with a gravimeter. See *gravimeter.*

gravitometer *n:* a device for measuring and recording the density or specific gravity of a gas or liquid passing a point of measurement. Also called a densimeter.

graywacke *n:* a sandstone that contains more than 15% silt and clay, and whose grains tend to be angular and poorly sorted.

growth fault *n:* an active fault that continues to slip while sediments are being deposited, causing the strata on the downthrust side to be thicker than those on the other side. Also called rollover fault.

guard-electrode log *n:* a focused system designed to measure the true formation resistivity in wellbores filled with salty mud.

gypsum *n:* a naturally occurring crystalline form of calcium sulfate in which each molecule of calcium sulfate is combined with two molecules of water.

H

half-life *n:* the amount of time needed for half of a quantity of radioactive substance to decay or transmute into a nonradioactive substance. Half-lives range from fractions of seconds to millions of years.

halite *n:* rock salt (NaCl).

hanging wall *n:* the rock surface forming the upper side of a fault when the fault plane is not vertical—that is, if the dip is less than 90°. Compare *footwall.*

hexane *n:* a liquid hydrocarbon of the paraffin series, C_6H_{14}.

horst *n:* a block of the earth's crust that has been raised up (relatively) between two faults; the opposite of a graben.

hydraulic fracturing *n:* an operation in which a specially blended liquid is pumped down a well and into a formation under pressure high enough to cause the formation to crack open, forming passages through which oil can flow into the wellbore. See *formation fracturing.*

hydrocarbons *n pl:* organic compounds of hydrogen and carbon, whose densities, boiling points, and freezing points increase as their molecular weights increase. Petroleum is a mixture of many different hydrocarbons.

hydrodynamic trap *n:* a petroleum trap in which the major trapping mechanism is the force of moving water.

hydrolysis *n:* the breaking down of a mineral by chemical reaction with water.

hydrostatic pressure *n:* the force exerted by a body of fluid at rest, which increases directly with the density and the depth of the fluid and is expressed in psi or kPa.

I

igneous rock *n:* a rock mass formed by the solidification of magma within the earth's crust or on its surface. Granite is an igneous rock.

imaging radar *n:* radar carried on airplanes or orbital vehicles that forms images of the terrain.

imbrication *n:* in sedimentary rocks, the arrangement of pebbles in a flat, overlapping pattern like bricks in a wall. Often shown by stream gravel deposits.

impermeable *adj:* preventing the passage of fluid. A formation may be porous yet impermeable if there is an absence of connecting passages between the voids within it. See *permeability.*

improved recovery *n:* the introduction of artificial drive and displacement mechanisms into a reservoir in order to produce a portion of the oil unrecoverable by primary recovery methods. See *enhanced oil recovery.*

induction log *n:* also called induction survey. See *induction survey.*

induction survey *n:* an electric well log in which the conductivity of the formation rather than the resistivity is measured.

infill drilling *n:* drilling wells between known producing wells to exploit the resources of a field to best advantage.

inorganic compounds *n pl:* chemical compounds that do not contain carbon as the principal element (excepting that in the form of carbonates, cyanides, and cyanates). Such compounds make up matter that is not plant or animal.

intangible development cost *n:* expense of an item that does not have a salvage value, such as site costs, rig transportation, rig operation, drilling fluid, formation tests, cement, well supplies, and other expenses relating to activities on a drilling rig.

interstice *n:* a pore space in a reservoir rock.

interstitial water *n:* water contained in the interstices, or pores, of reservoir rock. In reservoir engineering, it is synonymous with connate water. Compare *connate water.*

intrusive rock *n:* an igneous rock that, while molten, penetrated into or between other rocks and solidified.

ion *n:* an atom or a group of atoms charged either positively (a cation) or negatively (an anion) as a result of losing or gaining electrons.

isochore map *n:* a map on which points of equal drilled thickness of a formation are shown as a series of contours. Similar to an isopach map.

isolith map *n:* a map of a formation on which points of similar lithology are connected by a series of contours.

isometric diagram *n:* a drawing of a three-dimensional object in which lines parallel to the edges are drawn to scale without perspective or foreshortening.

isopach map *n:* a geological map of subsurface strata showing the various thicknesses of a given formation as a series of contours. It is widely used in calculating reserves and in planning improved recovery projects.

isopachous line *n:* a contour line drawn on a map joining points of equal thickness in a stratigraphic unit.

isostasy *n:* equilibrium between large segments of the earth's crust, which "float" on the denser mantle in such a way that thicker segments extend higher and deeper than thinner segments, and lighter blocks rise higher than denser blocks.

isotope *n:* a form of an element that has the same atomic number as its other forms but has a different atomic mass. Isotopes of an element have the same number of protons but different numbers of neutrons in the nucleus.

J

joint *n:* a crack or fissure produced in a rock by internal stresses.

K

kaolinite *n:* a light-colored clay mineral $[Al_2Si_2O_5(OH)_4]$.

L

lacustrine delta *n:* collection of sediment in a lake at the point at which a river or stream enters. When the flowing water enters the lake, the encounter with still water absorbs most or all of the stream's energy, causing its sediment load to be deposited.

laminar flow *n:* a smooth flow of fluid in which no turbulence or cross flow of fluid particles occurs between adjacent stream lines.

landform *n:* a recognizable, naturally formed, physical land feature having a characteristic shape, such as a plain, alluvial fan, valley, hill, or mountain.

landman *n:* a person in the petroleum industry who negotiates with landowners for land options, oil drilling leases, and royalties and with producers for the pooling of production in a field; also called a leaseman.

Landsat *n:* an unmanned earth-orbiting NASA satellite that transmits multispectral images to earth receiving stations; formerly called ERTS (Earth Resource Technology Satellite).

lateral focus log *n:* a resistivity log taken with a sonde that focuses an electrical current laterally, away from the wellbore, and into the formation being logged. Also called laterolog.

laterolog *n:* trade name for a Schlumberger guard or lateral focus resistivity log, but so commonly used as to be almost a generic term.

Laurasia *n:* the northern part of the supercontinent Pangaea, comprising the future land masses of North America, Greenland, and Eurasia.

lava *n:* magma that reaches the surface of the earth.

leaching *n:* in geology, the removal of minerals from rock by solution in water or another solvent.

lease *n:* 1. a legal document executed between a landowner, as lessor, and a company or individual, as lessee, that grants the right to exploit the premises for minerals or other products. 2. the area where production wells, stock tanks, separators, LACT units, and other production equipment are located.

lens *n:* 1. a porous, permeable, irregularly shaped sedimentary deposit surrounded by impervious rock. 2. a lenticular sedimentary bed that pinches out, or comes to an end, in all directions.

lessee *n:* the recipient of a lease (such as an oil and gas lease). Also called leasee.

lessor *n:* the conveyor of a lease (such as an oil and gas lease).

levee *n:* 1. an embankment that lies along the sides of a sea channel, a canyon, or a valley. 2. the low ridge sometimes deposited by a stream along its sides.

limestone *n:* a sedimentary rock rich in calcium carbonate; sometimes serves as a reservoir rock for petroleum.

lineament *n:* a linear topographic or tonal feature on the terrain and on images and maps of the terrain, thought to indicate a zone of subsurface structural weakness.

lithification *n:* the conversion of unconsolidated deposits into solid rock.

lithofacies map *n:* a facies map showing lithologic variations within a formation. It shows the variations of selected lithologic characteristics within a stratigraphic unit.

lithology *n:* 1. the study of rocks, usually macroscopic. 2. the individual character of a rock in terms of mineral composition, structure, and so forth.

loess *n:* unstratified, homogeneous accumulation of silt, often containing small amounts of clay or sand, redeposited by wind from glacial outwash or deserts.

log *n:* a systematic recording of data, such as a driller's log, mud log, electrical well log, or radioactivity log. Many different logs are run in wells to obtain various characteristics of downhole formations. *v:* to record data.

log a well *v:* to run any of the various logs used to ascertain downhole information about a well.

longshore current *n:* movement of seawater parallel to the shore.

M

magma *n:* the hot fluid matter within the earth's crust that is capable of intrusion or extrusion and that produces igneous rock when cooled.

magnetometer *n:* an instrument used to measure the intensity and direction of a magnetic field, especially that of the earth.

mantle *n:* the hot, plastic part of the earth that lies between the core and the crust. It begins 5–30 miles (10–50 km) beneath the surface and extends to 1,800 miles (2 900 km).

marine delta *n:* a sea-level extension of land shaped like the triangular Greek letter Δ; a depositional environment in which river-borne sediments accumulate as the flow energy of the river is dissipated in the ocean.

material balance *n:* a calculation used to inventory the fluids produced from a reservoir and the fluids remaining in a reservoir.

matrix *n:* in rock, the fine-grained material between larger grains, in which the larger grains are embedded. A rock matrix may be composed of fine sediments, crystals, clay, or other substances.

matrix acidizing *n:* the procedure by which acid flow is confined to the natural permeability and porosity of the formation, as opposed to fracture acidizing.

maximum efficiency rate *n:* the producing rate of a well that brings about maximum volumetric recovery from a reservoir with a minimum of residual-oil saturation at the time of depletion.

MER *abbr:* maximum efficiency rate.

Mesozoic era *n:* a span from 230 to 65 million years ago, the era of the dinosaurs and the first mammals.

metamorphic rock *n:* a rock derived from preexisting rocks by mineralogical, chemical, and structural alterations caused by heat and pressure within the earth's crust. Marble is a metamorphic rock.

metamorphism *n:* the process in which rock may be changed by heat and pressure into different forms.

methane *n:* a light, gaseous, flammable paraffinic hydrocarbon, CH_4, that has a boiling point of $-258°$ F and is the chief component of natural gas.

mica *n:* a silicate mineral characterized by sheet cleavage. Biotite is ferromagnesian black mica, and muscovite is potassic white mica.

micropaleontology *n:* paleontology dealing with fossils of microscopic size.

microresistivity log *n:* a resistivity logging tool consisting of a spring device and a pad. While the spring device holds the pad firmly against the borehole sidewall, electrodes in the pad measure resistivities in mud cake and nearby formation rock. See *resistivity well logging.*

millidarcy *n:* one-thousandth of a darcy.

mineral *n:* 1. a naturally occurring inorganic crystalline element or compound. It has a definite chemical composition and characteristic physical properties such as crystal shape, melting point, color, and hardness. Most minerals found in rocks are not pure. 2. broadly, a naturally occurring homogeneous substance that is obtained from the ground for man's use (e.g., stone, coal, salt, sulfur, sand, petroleum, water, natural gas).

molasse *n:* a thick sequence of soft, ungraded, cross-bedded, fossiliferous marine and terrestrial conglomerates, sandstones, and shales derived from the erosion of growing mountain ranges.

molecule *n:* the smallest particle of a substance that retains the properties of the substance; it is composed of one or more atoms.

mud analysis *n:* examination and testing of drilling mud to determine its physical and chemical properties.

mud logging *n:* the recording of information derived from examination and analysis of formation cuttings made by the bit and of mud circulated out of the hole.

mudstone *n:* 1. a massive, blocky rock composed of approximately equal proportions of clay and silt, lacking the fine lamination of shale. 2. generally, rock consisting of an indefinite and variable mixture of clay, silt, and sand particles.

multiple completion *n:* an arrangement for producing a well in which one wellbore penetrates two or more petroleum-bearing formations.

N

naphthene series *n:* the saturated hydrocarbon compounds of the general formula C_nH_{2n} (e.g., ethene or ethylene, C_2H_4). They are cycloparaffin derivatives of cyclopentane (C_5H_{10}) or cyclohexane (C_6H_{12}) found in crude petroleum.

nappe *n:* a large body of rock that has been thrust horizontally over neighboring rocks by compressive forces, as during the collision of two continents.

neutron *n:* a part of the nucleus of all atoms except hydrogen. Under certain conditions neutrons can be emitted from a substance when its nucleus is penetrated by gamma particles from a highly radioactive source. This phenomenon is used in neutron logging.

neutron log *n:* a type of radioactivity well logging in which the rock around the wellbore is bombarded with radiation, and the amount of radiation that is not absorbed is recorded. See *radioactivity well logging.*

nonassociated gas *n:* gas in a reservoir that contains no oil.

nonconformity *n:* a buried landscape in which sediments were deposited on an eroded surface of igneous or metamorphic rock.

nonfoliated metamorphic rock *n:* metamorphic rock that appears massive and homogeneous, without the layered look of foliated metamorphics.

normal fault *n:* a dip-slip fault along which the hanging wall has subsided relative to the footwall.

O

oblique slip *n:* slip at an angle between the dip and the strike in a fault plane.

obsidian *n:* an extrusive igneous rock having cooled so rapidly that no crystals formed at all; volcanic glass.

octane *n:* a paraffinic hydrocarbon, C_8H_{18}, a liquid at atmospheric conditions, with a boiling point of 258° F (at 14.7 psi).

offset well *n:* a well drilled on a tract of land next to another owner's tract on which there is a producing well.

oil shale *n:* a shale containing hydrocarbons that cannot be recovered by an ordinary oilwell but can be extracted by mining and processing.

oil-water contact *n:* the plane (typically a zone several feet thick) at which the bottom of an oil sand contacts the top of a water sand in a reservoir; the oil-water interface.

oil-wet reservoir *n:* a hydrocarbon reservoir in which the grains of rock are coated not with water but with oil; rare.

oil window *n:* See *petroleum window.*

oolite *n:* an ovoid, sandlike particle that is formed when calcite accretes on a smaller particle.

open-hole completion *n:* a method of preparing a well for production in which no production casing or liner is set opposite the producing formation. Also called a barefoot completion.

Ordovician *adj:* of or relating to the geologic period from approximately 500 million to 430 million years ago during the early part of the Paleozoic era, or to the rocks formed during this period.

organic compounds *n pl:* chemical compounds that contain carbon atoms, either in straight chains or in rings, and hydrogen atoms. They may also contain oxygen, nitrogen, or other atoms.

orthoclase *n:* a light-colored feldspar mineral ($KAlSi_3O_8$), common in granite.

outcrop *n:* part of a formation exposed at the earth's surface. *v:* to appear on the earth's surface (as a rock).

outwash *n:* sediment deposited by meltwater streams beyond an active glacier.

overburden *n:* the strata of rock that lie above the stratum of interest in drilling.

overriding royalty *n:* an interest carved out of the lessee's working interest, entitling its owner to a fraction of production free of any production or operating expense, but not free of production or severance tax levied on production.

overthrust fault *n:* a low-dip angle (nearly horizontal) reverse fault along which a large displacement has occurred. Some overthrusts, such as many of those in the Rocky Mountain Overthrust Belt, represent slippages of many miles.

overturned fold *n:* a rock fold that has become slanted to one side so that the layers on one side appear to occur in reverse order (younger layers beneath older).

oxidation *n:* a chemical reaction in which a compound loses electrons and gains a more positive charge.

oxidation-reduction potential *n:* the difference in voltage shown when an inert electrode is immersed in a reversible oxidation-reduction system; it is the measurement of the system's state of oxidation. Also called Eh, ORP, or redox potential.

oxidize *v:* 1. to combine with oxygen. 2. to remove one or more electrons from (an atom, ion, or molecule).

oxygenation *n:* combining or supplying with oxygen

P

packer *n:* a piece of downhole equipment, consisting of a sealing device, a holding or setting device, and an inside passage for fluids, used to block the flow of fluids through the annular space between the tubing and the wall of the wellbore by sealing off the space between them.

paleo- *comb form:* ancient; early; long ago; primitive.

paleogeography *n:* geography of a specified geologic past.

paleontology *n:* a science that concerns the life of past geologic periods, studying especially fossil forms and the chronology of the earth.

Paleozoic era *n:* a span of time from 600 million to 230 million years ago, during which developed a great diversification of life forms.

panel diagram *n:* a diagram of a block of earth in which a series of cross sections are joined together and viewed obliquely from above; useful in showing how formation structure and stratigraphic thickness vary both horizontally and vertically. Also called fence diagram.

Pangaea *n:* the supercontinent comprising all the principal continental masses near the beginning of the Mesozoic era.

paraffin *n:* a saturated aliphatic hydrocarbon having the formula C_nH_{2n+2} (e.g., methane, CH_4; ethane, C_2H_6). Heavier paraffin hydrocarbons (i.e., $C_{18}H_{38}$ and heavier) form a waxlike substance that is called paraffin.

pay *n:* See *pay sand.*

pay sand *n:* the producing formation, often one that is not even sandstone. It is also called pay, pay zone, and producing zone.

pay zone *n:* See *pay sand.*

peat *n:* an organic material that forms by the partial decomposition and disintegration of vegetation in tropical swamps and other wet, humid areas. It is the precursor of coal.

peg model *n:* an analog model of three dimensions used to study the structure and stratigraphy of a subsurface area; it is made by placing pegs of varying heights into a flat platform to represent the structural contours of strata.

Pennsylvanian period *n:* a geologic time period in the Paleozoic era, from 320 to 280 million years ago. Also, the latter part of the Carboniferous period. It was named for the outcrops of coal in Pennsylvania.

pentane *n:* a liquid hydrocarbon of the paraffin series, C_5H_{12}.

perforate *v:* to pierce the casing wall and cement of a wellbore to provide holes through which formation fluids may enter or to provide holes in the casing so that materials may be introduced into the annulus between the casing and the wall of the borehole.

perforation *n:* a hole made in the casing, cement, and formation, through which formation fluids enter a wellbore.

permeability *n:* 1. a measure of the ease with which a fluid flows through the connecting pore spaces of rock or cement. The unit of measurement is the millidarcy. 2. fluid conductivity of a porous medium. 3. ability of a fluid to flow within the interconnected pore network of a porous medium. See *absolute permeability, effective permeability,* and *relative permeability.*

Permian period *n:* the last geologic time period in the Paleozoic era, 280 to 225 million years ago.

petroleum *n:* a substance occurring naturally in the earth in solid, liquid, or gaseous state and composed mainly of mixtures of chemical compounds of carbon and hydrogen, with or without other nonmetallic elements such as sulfur, oxygen, and nitrogen.

petroleum geology *n:* the study of oil- and gas-bearing rock formations. It deals with the origin, occurrence, movement, and accumulation of hydrocarbon fuels.

petroleum window *n:* the conditions of temperature and pressure under which petroleum will form. Also called oil window.

petrology *n:* a branch of geology dealing with the origin, occurrence, structure, and history of rocks, principally igneous and metamorphic rocks. Compare *lithology.*

phosphate *n:* generic term for any compound containing a phosphate group (PO_4^-); a salt or ester of phosphoric acid.

phosphorite *n:* a rock of biochemical origin, composed largely of calcium phosphate from bird droppings and vertebrate remains.

photosynthesis *n:* a process by which chlorophyll-bearing plants produce simple sugars from carbon dioxide and water using the energy of sunlight.

piercement dome *n:* See *diapir.*

pinchout *n:* an oil-bearing stratum that forms a trap for oil and gas by narrowing and tapering off within an impervious formation.

plagioclase *n:* a common rock-forming mineral varying in composition from sodium aluminum silicate ($NaAlSi_3O_8$) to calcium aluminum silicate ($CaAl_2Si_2O_8$).

planimeter *n:* an instrument for measuring the area of a plane figure. As the point on a tracing arm is passed along the outline of a figure, a graduated wheel and disk indicate the area encompassed.

plate tectonics *n:* movement of great crustal plates of the earth upon slow currents in the plastic mantle, similar to the movement of boxes on a conveyor belt. Today geologists believe the the earth's crust is divided into six major plates and several smaller ones atop some of which the continents are carried away from a system of midocean ridges and toward another system of deep-sea trenches. Compare *continental drift.*

playa *n:* the flat bottom of an undrained desert basin that at times becomes a shallow lake. Also called sebkha or sabkha.

Pleistocene *adj:* of or relating to the geologic epoch from about 1 million to 10 thousand years ago, the first part of the Quaternary period of the Cenozoic era, sometimes called the Ice Age, which extended from the end of the Tertiary period until the last retreat of the northern continental ice sheets; of or relating to the rocks or sediments formed during this epoch.

plug and abandon *v:* to place cement plugs into a dry hole and abandon it.

plunging fold *n:* a fold of rock whose long axis is not horizontal.

pluton *n:* a large subterranean body of igneous rock.

polymer *n:* a substance that consists of large molecules formed from smaller molecules in repeating structural units (monomers).

pool *n:* a reservoir or group of reservoirs. The term is a misnomer in that hydrocarbons seldom exist in pools but rather in the pores of rock. *v:* to combine small or irregular tracts into a unit large enough to meet state spacing regulations for drilling.

pooling *n:* the combining of small or irregular tracts into a unit large enough to meet state spacing regulations for drilling. Compare *unitization.*

pore *n:* an opening or space within a rock or mass of rocks, usually small and often filled with some fluid (water, oil, gas, or all three). Compare *vug.*

porosity *n:* 1. the condition of being porous (such as a rock formation). 2. the ratio of the volume of empty space to the volume of solid rock in a formation, indicating how much fluid a rock can hold.

porous *adj:* having pores, or tiny openings, as in rock.

potash *n:* potassium carbonate (K_2CO_3).

potentiometric surface *n:* a surface representing the hydrodynamic pressure gradient of groundwater flowing through an aquifer; the level to which unconfined flowing water would rise.

Precambrian era *n:* a span of 4 billion years from the earth's beginning until 600 million years ago, during which the earth was devoid of all but the most primitive life forms.

precipitation *n:* the production of a separate liquid phase from a mixture of gases, as rain, or of a separate solid phase from a liquid solution, as in the precipitation of calcite cement from water in the interstices of rock.

pressure buildup plot *n:* a logarithmic plot of bottomhole buildup pressure versus time.

pressure differential *n:* also called differential pressure. See *differential pressure.*

primary migration *n:* movement of hydrocarbons out of source rock into reservoir rock.

primary term *n:* the specified duration of an oil and gas lease (e.g., 3 years), within which time a well must be drilled in order to keep the lease in effect.

producing zone *n:* the zone or formation from which oil or gas is produced. See *pay sand.*

production *n:* 1. the phase of the petroleum industry that deals with bringing the well fluids to the surface and separating them and with storing, gauging, and otherwise preparing the product for the pipeline. 2. the amount of oil or gas produced in a given period.

production tax *n:* a state or municipal tax on oil and gas products, levied at the wellhead for the removal of the hydrocarbons. Also called severance tax.

progradation *n:* the seaward buildup of a beach, delta, or fan by nearshore deposition of sediments by a river, by waves, or by longshore currents.

propane *n:* a paraffinic hydrocarbon, (C_3H_8), that is a gas at ordinary atmospheric conditions but is easily liquefied under pressure. It is a constituent of liquefied petroleum gas.

proppant *n:* also called propping agent. See *propping agent.*

propping agent *n:* a granular substance (sand grains, aluminum pellets, or other material) that is carried in suspension by the fracturing fluid and that serves to keep the cracks open when fracturing fluid is withdrawn after a fracture treatment.

proration *n:* a system, enforced by a state or federal agency or by agreement between operators, that limits the amount of petroleum that can be produced from a well or a field within a given period.

prospect *n:* 1. an area of land under exploration that has good possibilities of producing profitable minerals. 2. the set of circumstances, both geologic and economic, that justify drilling a wildcat well. *v:* to examine the surface and subsurface of an area of land for signs of mineral deposits.

proton *n:* the positively charged elementary particle of an atomic nucleus, occurring in the nucleus along with the neutron.

proved reserves of crude oil *n pl:* according to API standard definitions, proved reserves of crude oil as of December 31 of any given year are the estimated quantities of all liquids statistically defined as crude oil which geological and engineering data demonstrate with reasonable certainty to be recoverable in future years from known reservoirs under existing economic and operating conditions.

proved reserves of natural gas *n pl:* according to API standard definitions, proved reserves of natural gas as of December 31 of any given year are the estimated quantities of natural gas which geological and engineering data demonstrate with reasonable certainty to be recoverable in the future from known natural oil and gas reservoirs under existing economic and operating conditions.

psi *abbr:* pounds per square inch.

pumice *n:* vesicular obsidian formed from gas-filled lava that cooled rapidly; it is often light enough to float on water.

pyroclastic particles *n pl:* particles produced directly by volcanic action when gases within molten lava expand rapidly and the water suddenly flashes into steam, blasting the molten mass into tiny splinters of solidifying glass. The hot particles eventually come to rest in thick blankets of cooling cinders, called ash.

pyroclastic rock *n:* rock formed from pyroclastic particles. See *pyroclastic particles.*

Q

quartz *n:* a hard mineral composed of silicon dioxide (silica); a common component in igneous, metamorphic, and sedimentary rocks.

R

radioactive decay *n:* the spontaneous transformation of a radioactive atom into one or more different atoms or particles, resulting in a long-term transformation of the radioactive element into lighter, nonradioactive elements.

radioactivity *n:* the property possessed by some substances (such as radium, uranium, and thorium) of releasing alpha particles, beta particles, or gamma rays as the substance spontaneously disintegrates.

radioactivity log *n:* a record of the natural or induced radioactive characteristics of subsurface formations. See *radioactivity well logging.*

radioactivity well logging *n:* the recording of the natural or induced radioactive characteristics of subsurface formations.

radioisotope *n:* an element, or one of its variants, that exhibits radioactivity.

radiometric *adj:* relating to the measurement of geologic time by means of the rate at which radioactive elements disintegrate.

radiometric dating *n:* a technique for measuring the age of an object or a sample of material by determining the ratio of the concentration of a radioisotope to that of a stable isotope in it.

recumbent fold *n:* a fold of rock in which the axial plane of an overturned fold has become horizontal or nearly so.

red bed *n:* a layer of sedimentary rock that is predominantly red, especially one of Permian or Triassic age.

reduction *n:* adding one or more electrons to (an atom or ion or molecule).

reef *n:* 1. a type of reservoir trap composed of rock (usually limestone) formed from the shells or skeletons of marine animals. 2. a buried coral or other reef from which hydrocarbons may be withdrawn.

refraction *n:* deflection from a straight path undergone by a light ray or energy wave in passing from one medium to another in which the wave velocity is different, such as the bending of light rays when passing from air into water.

regional metamorphism *n:* a type of metamorphism that occurs in bodies of rock that have been deeply buried or greatly deformed by tectonic changes.

relative permeability *n:* the ratio of effective permeability to absolute permeability. The relative permeability of rock to a single fluid is 1.0 when only that fluid is present, and 0.0 when the presence of another fluid prevents all flow of the given fluid. Compare *absolute permeability, effective permeability.*

reserves *n pl:* the unproduced but recoverable oil or gas in a formation that has been proved by production.

reservoir *n:* a subsurface, porous, permeable rock body in which oil and/or gas has accumulated.

reservoir rock *n:* a permeable rock that may contain oil or gas in appreciable quantity and through which petroleum may migrate.

reservoir simulation *n:* model of a reservoir done by computer for the purpose of predicting reservoir behavior and showing production over time, allowing decisions about managing the reservoir to be made.

resistivity *n:* the electrical resistance offered to the passage of current; the opposite of conductivity.

resistivity well logging *n:* the recording of the resistance of formation water to natural or induced electrical current. See *electric well log.*

reverse fault *n:* a dip-slip fault along which the hanging wall has moved upward relative to the footwall. Also called a thrust fault.

rhyolite *n:* a light-colored, fine-grained volcanic rock, the extrusive equivalent of granite.

rift zone *n:* the zone along which crustal plates separate because of slowly diverging convection currents in the semisolid, deformable mantle.

risk analysis *n:* in the oil business, the activity of assigning probabilities to the expected outcomes of a drilling venture.

rock *n:* an indurated aggregate of different minerals. Rocks are divided into three groups on the basis of their mode of origin: igneous, metamorphic, and sedimentary.

rock cycle *n:* the possible sequences of events, all interrelated, by which rocks may be formed, changed, destroyed, or transformed into other types of rocks. The events include formation from magma, erosion, sedimentation, and metamorphism.

rock stratigraphic unit *n:* a distinctive body of rock that can be identified by its lithologic or structural features regardless of its fossils or time boundaries; commonly, a formation. Compare *time stratigraphic unit.*

rock texture *n:* all of the properties relating to the grain-to-grain relationships of a rock. Textural properties include chemical composition, grain shape and roundness, grain size and sorting, grain orientation, porosity, and permeability. See *clastic texture* and *crystalline texture.*

rollover anticline *n:* an anticline formed when the dip of a growth fault approaches the horizontal at depth, and deposition is faster on the downthrown side, which tends to "roll over" or curl downward.

rollover fault *n:* also called growth fault. See *growth fault.*

royalty *n:* the portion of oil, gas, and minerals retained by the lessor upon execution of a lease; or their cash value paid by the lessee to the lessor or to one who has acquired possession of the royalty rights, based on a certain percentage of the gross production from the property.

S

sample log *n:* a graphic representative model of the rock formations penetrated by drilling, prepared by the geologist from samples and cores.

samples *n pl:* 1. the well cuttings obtained at designated footage intervals during drilling. From an examination of these cuttings, the geologist determines the type of rock and formations being drilled and estimates oil and gas content. 2. small quantities of well fluids obtained for analysis.

sandstone *n:* a sedimentary rock composed of individual mineral grains of rock fragments between $\frac{1}{16}$

and 2 mm in diameter and cemented together by silica, calcite, iron oxide, and so forth. Sandstone is commonly porous and permeable and therefore a likely type of rock in which to find a petroleum reservoir.

saturated hydrocarbons *n pl:* hydrocarbon compounds in which all carbon valence bonds are filled with hydrogen atoms. Natural gas and natural gas liquids are saturated compounds.

scarp *n:* an extended cliff or steep slope separating two level or gently sloping areas, produced by erosion or faulting; an escarpment.

schist *n:* a coarse-grained, foliated metamorphic rock that splits easily into layers, formed when shale under deep burial becomes slate and then with more intense metamorphism becomes schist.

screen liner *n:* a pipe that is perforated and arranged with a wire wrapping to act as a sieve to prevent or minimize the entry of sand particles into the wellbore. Also called a screen pipe.

sealing fault *n:* a fault that contains material of low permeability, such as gouge.

sebkha (or **sabkha**) *n:* See *playa.*

secondary migration *n:* movement of hydrocarbons, subsequent to primary migration, through porous, permeable reservoir rock, by which oil and gas become concentrated in one locality.

secondary porosity *n:* porosity created in formation after it has formed, either because of dissolution or stress distortion taking place naturally, or because of treatment by acid or injection of coarse sand.

sedimentary rock *n:* a rock composed of materials that were transported to their present position by wind or water. Sandstone, shale, and limestone are sedimentary rocks.

sedimentation *n:* the process of deposition of layers of clastic particles or minerals that settle out of water, ice, or other transporting media.

sedimentology *n:* the science dealing with the description, classification, and interpretation of sediments and sedimentary rock.

seismic *adj:* of or relating to an earthquake or earth vibration, including those artificially induced.

seismic data *n:* detailed information obtained from earth vibration produced naturally or artificially (as in geophysical prospecting).

seismograph *n:* a device that detects vibrations in the earth, used in studying the earth's interior and in prospecting for probable oil-bearing structures.

seismology *n:* the study of earth vibrations.

severance tax *n:* also called production tax. See *production tax.*

shale *n:* a fine-grained sedimentary rock composed mostly of consolidated clay or mud. Shale is the most frequently occurring sedimentary rock.

shoe *n:* a device placed at the end of or beneath an object for various purposes (e.g., casing shoe, guide shoe).

shoestring sand *n:* a narrow, often sinuous sand deposit, usually a buried sandbar or filled channel.

shoreface *n:* that part of the seashore seaward of the low-tide mark that is affected by wave action.

shut in *v:* 1. to close the valves on a well so that it stops producing. 2. to close in a well in which a kick has occurred.

shut-in bottomhole pressure *n:* the pressure at the bottom of a well when the surface valves on the well are completely closed, caused by formation fluids at the bottom of the well.

sidewall coring *n:* a coring technique in which core samples are obtained from the hole wall in a zone that has already been drilled.

silica *n:* a mineral that has the chemical formula SiO_2 (silicon dioxide). It is relatively hard and insoluble. Quartz is a form of silica, but usually contains impurities that give it color.

silicate *n:* generally, a compound that contains silicon and oxygen.

siliceous *adj:* containing abundant silica, or silicon dioxide (SiO_2).

siltstone *n:* a fine-grained, shalelike sedimentary rock composed mostly of particles $\frac{1}{16}$ to $\frac{1}{256}$ mm in diameter.

skin *n:* 1. the area of the formation that is damaged because of the invasion of foreign substances into the exposed section of the formation adjacent to the wellbore during drilling and completion. 2. the pressure drop from the outer limits of drainage to the wellbore caused by the relatively thin veneer (or skin) of the affected formation.

slate *n:* a metamorphic rock formed when shale becomes buried deeply; the heat and pressure fuse individual mineral grains into slate.

solution gas *n:* lighter hydrocarbons that exist as a liquid under reservoir conditions but that effervesce as gas when pressure is released during production.

sonar *n:* an apparatus that detects the presence of an underwater object by sending out sonic or supersonic waves that are reflected back to it by the object.

sonde *n:* a logging tool assembly, especially the device in the logging assembly that senses and transmits formation data.

sonic logging *n:* See *acoustic well logging.*

source rock *n:* rock within which oil or gas is generated from organic materials.

spontaneous potential *n:* one of the natural electrical characteristics exhibited by a formation as measured by a logging tool lowered into the wellbore.

Also referred to as self-potential, it is one of the basic curves obtained by an electrical well log; usually referred to by the initials *SP.*

spontaneous potential curve *n:* a measurement of the electrical currents that occur in the wellbore when fluids of different salinities are in contact. Also called self-potential curve.

spontaneous potential log *n:* a record of a spontaneous potential curve.

step-out well *n:* a well drilled adjacent to or near a proven well to ascertain the limits of the reservoir; an outpost well.

strata *n pl:* distinct, usually parallel, and originally horizontal beds of rock. An individual bed is a stratum.

stratigraphic test *n:* a borehole drilled primarily to gather information on rock types and sequence.

stratigraphic trap *n:* a petroleum trap that occurs when the top of the reservoir bed is terminated by other beds or by a change of porosity or permeability within the reservoir itself. Compare *structural trap.*

stratigraphy *n:* a branch of geology concerned with the study of the origin, composition, distribution, and succession of rock strata.

strat test *n:* common term for stratigraphic test.

stratum *n:* See *strata.*

strike *n:* See *formation strike.*

strike slip *n:* horizontal displacement along a fault plane; the San Andreas fault in California is a strike-slip fault.

stringer *n:* a relatively narrow splinter of a rock formation that is stratigraphically disjoint, interrupting the consistency of another formation and making drilling that formation less predictable.

structural trap *n:* a petroleum trap that is formed because of deformation (as folding or faulting) of the reservoir formation. Compare *stratigraphic trap.*

subcrop *n:* the area within which a formation occurs directly beneath an unconformity.

subduction zone *n:* a deep trench formed in the ocean floor along the line of convergence of oceanic crust with other oceanic or continental crust when one plate (always oceanic) dives beneath the other. The plate that descends into the hot mantle is partially melted. Magma rises through fissures in the heavier, unmelted crust above, creating a line of plutons and volcanoes that eventually form an island arc parallel to the trench.

subgeologic map *n:* a map of the formations directly above an unconformity. Also called a worm's-eye map.

sulfate *n:* a compound containing the SO_4^{--} group, as in calcium sulfate $(CaSO_4)$.

superposition *n:* 1. the order in which sedimentary layers are deposited, with the oldest layer on bottom, the youngest layer on top. 2. the process of sedimentary layering.

supersaturation *n:* the condition of containing more solute in solution than would normally be present at the existing temperature.

support agreement *n:* an agreement between petroleum companies, in which one contributes money or acreage to another's drilling operation in return for information gained from the drilling.

surface casing *n:* also called surface pipe. See *surface pipe.*

surface pipe *n:* the first string of casing (after the conductor pipe) that is set in a well, varying in length from a few hundred to several thousand feet. Some states require a minimum length to protect freshwater sands. Compare *conductor pipe.*

surface tension *n:* the tendency of liquids to maintain as small a surface as possible, caused by the cohesive attraction between the molecules of liquid.

suspended load *n:* in a flowing stream of water, the finer sand, silt, and clay that are carried well off the bottom by the turbulence of the water. Compare *bed load* and *dissolved load.*

syncline *n:* a trough-shaped configuration of folded rock layers. Compare *anticline.*

T

talus *n:* angular pieces of rock produced by weathering that come to rest in a steep slope at the bottom of a mountainside.

tar sand *n:* a sandstone that contains chiefly heavy, tarlike hydrocarbons.

tectonic *adj:* or or relating to the deformation of the earth's crust, the forces involved in or producing such deformation, and the resulting rock forms.

Tethys Sea *n:* an ancient great ocean of the Cretaceous period, 135 to 65 million years ago, between Eurasia and Africa.

thorium *n:* a radioactive metallic element found combined in minerals.

thrust fault *n:* See *reverse fault.*

thumper *n:* a hydraulically operated hammer used in obtaining a seismograph in oil exploration. It is mounted on a vehicle and, when dropped, creates shock waves in the subsurface formations, which are recorded and interpreted to reveal geological information.

time stratigraphic unit *n:* a layer of rock, with or without facies variations, deposited during a distinct geologic time interval. Compare *rock stratigraphic unit.*

topography *n:* the configuration of a land surface, including its natural and man-made features with their relative positions and elevations.

topset bed *n:* a part of a marine delta that is nearest the shore and that is composed of the heavier, coarser particles carried by the river.

transform fault *n:* a strike-slip fault caused by relative movement between crustal plates.

trap *n:* a body of permeable oil-bearing rock surrounded or overlain by an impermeable barrier that prevents oil from escaping.

trilobite *n:* one of a class of extinct Paleozoic arthropods.

turbidite *n:* a characteristic sedimentary deposit of the continental rise, formed by a turbidity current and composed of clay, silt, and gravel with the clay on top. See *turbidity current.*

turbidity current *n:* a dense mass of sediment-laden water that flows down the continental slope, typically through a submarine canyon.

U

unconformity *n:* 1. lack of continuity in deposition between rock strata in contact with one another, corresponding to a gap in the stratigraphic record. 2. the surface of contact between rock beds in which there is a discontinuity in the ages of rocks. See *angular unconformity* and *disconformity.*

uniformitarianism *n:* the geologic principle that the processes that are at work today on the earth are the same as, or very similar to, the processes that affected the earth in the past. Also called gradualism. Compare *catastrophism.*

United States Geological Survey *n:* a federal agency established in 1879 to conduct investigations of the geological structure, mineral resources, and products of the U.S.A., whose activities include assessing onshore and offshore mineral resources; providing information for society to mitigate the impact of floods, earthquakes, landslides, volcanoes, and droughts; monitoring the nation's groundwater and surface water supplies and man's impact thereon; and providing mapped information on the nation's landscape and land use.

unitization *n:* the combining of leased tracts on a field-wide or reservoir-wide scale so that many tracts may be treated as one to facilitate operations like secondary recovery. Compare *pooling.*

unsaturated hydrocarbon *n:* a straight-chain compound of hydrogen and carbon whose total combining power has not yet been reached and to which other atoms or radicals can be added.

updip *adj:* higher on the formation dip angle than a particular point.

USGS *abbr:* United States Geological Survey.

V

vesicular *adj:* containing vesicles, or small cavities.

viscous *adj:* having a high resistance to flow.

vug *n:* a cavity in a rock; a small cavern, larger than a pore but too small to contain a person. Typically found in limestone subject to leaching by groundwater.

W

water-wet reservoir *n:* a hydrocarbon reservoir whose rock grains are coated with a film of water.

weathering *n:* the breakdown of large rock masses into smaller pieces by physical and chemical climatological processes.

welded tuff *n:* a pyroclastic deposit hardened by the action of heat, pressure from overlying material, and hot gases.

well completion *n:* 1. the activities and methods of preparing a well for the production of oil and gas or for other purposes, such as injection; the method by which one or more flow paths for hydrocarbons are established between the reservoir and the surface. 2. the system of tubulars, packers, and other tools installed beneath the wellhead in the production casing; that is, the tool assembly that provides the hydrocarbon flow path or paths.

wellhead *n:* the equipment installed at the surface of the wellbore. A wellhead includes such equipment as the casinghead and tubing head. *adj:* pertaining to the wellhead (e.g., wellhead pressure).

well logging *n:* the recording of information about subsurface geologic formations, including records kept by the driller and records of mud and cutting analyses, core analysis, drill stem tests, and electric, acoustic, and radioactivity procedures.

well log library *n:* a private, or sometimes public, organization that maintains collections of oil field data, particularly well logs. An individual or company that uses the information gains access to it by paying membership dues or a user's fee.

well spacing *n:* regulation for conservation purposes of the number and location of wells over a reservoir.

well stimulation *n:* any of several operations used to increase the production of a well. See *acidize* and *formation fracturing.*

wildcat *n:* a well drilled in an area where no oil or gas production exists. *v:* to drill wildcat wells.

windfall profit tax *n:* a federal excise tax on crude oil, which has a different rate for oil in a number of categories, for example, newly discovered oil, stripper oil, stripper oil produced by independents, and so on.

wireline *n:* a small-diameter metal line used in wireline operations; also called slick line. Compare *conductor line.*

workover *n:* the performance of one or more of a variety of remedial operations on a producing oilwell to try to increase production. Examples of workover jobs are deepening, plugging back, pulling and resetting liners, squeeze cementing, and so forth.

worm's-eye map *n:* See *subgeologic map.*

BIBLIOGRAPHY

Amyx, James W., D. M. Bass, Jr., and R. L. Whiting. *Petroleum Reservoir Engineering.* New York: McGraw-Hill, 1960.

Beck, Frederik W. Mansvelt, and Karl M. Wiig. *The Economics of Offshore Oil and Gas Supplies.* Lexington, Mass.: Lexington Books, 1977.

Beckmann, H., ed. *Petroleum Engineering.* Vol. 3, *Geology of Petroleum.* Stuttgart: Ferdinand Enke Publishing, 1976.

Beebe, W., and Bruce F. Curtis, eds. *Natural Gases of North America.* 2 vols. Tulsa: American Association of Petroleum Geologists, 1968.

Brown, L. F., Jr., and W. L. Fisher. *Seismic Stratigraphic Interpretation and Petroleum Exploration.* Tulsa: American Association of Petroleum Geologists, 1979.

——. *Geology and Geometry of Depositional Systems.* Tulsa: American Association of Petroleum Geologists, 1979.

Burk, Creighton A., and Charles L. Drake. *Impact of the Geosciences on Critical Energy Resources.* Boulder, Colo.: American Association for the Advancement of Science, 1978.

Calder, Nigel. *The Restless Earth.* New York: Viking Press, 1972.

Carson, W. G. *The Other Price of Britain's Oil.* New Brunswick, N. J.: Rutgers University Press, 1982.

Case, L. C. *Water Problems in Oil Production.* Tulsa: Petroleum Publishing, 1970.

Chapman, Richard E. *Petroleum Geology.* New York: Elsevier Scientific Publishing, 1973.

Clark, James A. *Chronological History of the Petroleum and Natural Gas Industries.* Houston: Clark Book Company, 1963.

Clark, J. A., ed. *Ahead of His Time.* Houston: Gulf Publishing, 1971.

Clark, J. A., and Michel T. Halbouty. *The Last Boom.* New York: Random House, 1972.

Clark, N. J. *Elements of Petroleum Reservoirs.* Dallas: American Institute of Mining, Metallurgical, and Petroleum Engineers, 1969.

Coffeen, J. A. *Seismic Exploration Fundamentals.* Tulsa: Petroleum Publishing, 1978.

Continents Adrift—Readings from Scientific American. San Francisco: W. H. Freeman and Company, 1971.

Dickey, Parke A. *Petroleum Development Geology.* Tulsa: Petroleum Publishing, 1979.

Dickinson, William R., and H. Yarborough. *Plate Tectonics and Hydrocarbon Accumulation.* 4th ed. Tulsa: American Association of Petroleum Geologists, 1978.

Dix, C. H. *Seismic Prospecting for Oil.* Boston: International Human Resources Development, 1981.

Edwards, L. M., et al. *Handbook of Geothermal Energy.* Houston: Gulf Publishing, 1982.

Eicher, Don L., and A. Lee McAlester. *History of the Earth.* Englewood Cliffs, N. J.: Prentice-Hall, 1980.

Ellison, Samuel P. *General Geology Laboratory Workbook.* New York: Harper & Row, 1958.

Fairbridge, Rhodes W. "The Earth's Climate and Environment through 4.5 Billion Years." In *Rediscovery of the Earth*, edited by Lloyd Motz. New York: Van Nostrand Reinhold, 1975.

Fettweis, G. B. *World Coal Resources*. Amsterdam: Elsevier Scientific Publishing, 1979.

Flawn, Peter T. *Environmental Geology*. New York: Harper & Row, 1970.

Foster, Robert J. *Physical Geology*. Columbus: Charles E. Merrill Publishing, 1983.

Friedman, Gerald M., and John E. Sanders. *Principles of Sedimentology*. New York: John Wiley and Sons, 1978.

Garland, G. D. *Introduction to Geophysics*. Philadelphia: W. B. Saunders, 1971.

Garrison, Paul. *Investing in Oil in the '80s*. Tulsa: PennWell Books, 1981.

Gottlieb, Ben M. *Unconventional Methods in Exploration for Petroleum and Natural Gas*. Dallas: SMU Press, 1981.

Griffiths, D. H., and R. F. King. *Applied Geophysics for Geologists and Engineers*. Oxford: Pergamon Press, 1981.

Halbouty, Michel T. *Giant Oil and Gas Fields of the Decade 1968-1978*. Memoir #30. Tulsa: American Association of Petroleum Geologists, 1980.

———. *The Deliberate Search for the Subtle Trap*. Memoir #32. Tulsa: American Association of Petroleum Geologists, 1982.

Hamblin, W. Kenneth. *The Earth's Dynamic Systems*. Minneapolis: Burgess, 1975.

Haun, John T., ed. *Methods of Estimating the Volume of Undiscovered Oil and Gas Resources*. AAPG Studies in Geology No. 1. Tulsa: American Association of Petroleum Geologists, 1975.

———. *Origin of Petroleum: Selected Papers Reprinted from AAPG Bulletin, 1950-1969*. Tulsa: American Association of Petroleum Geologists, 1976.

Hedley, Don. *World Energy: The Facts and the Future*. London: Euromonitor Publications, 1981.

Herald, Frank A., ed. *Occurrence of Oil and Gas in West Texas*. Austin: The University of Texas, 1957.

Hilchie, Douglas W. *Applied Openhole Log Interpretation*. Golden, Colo.: Douglas W. Hilchie, 1978.

Hobson, G. D., ed. *Developments in Petroleum Geology*, vol. 1. London: Applied Science Publishers, 1977.

Hobson, G. D., and E. N. Tiratsoo. *Introduction to Petroleum Geology*. 2d ed. Houston: Gulf Publishing, 1981.

Hubbert, Marion King. *Energy Resources*. Washington: National Academy of Sciences, 1962.

———. *U. S. Energy Resources, A Review as of 1972*. Washington: Government Printing Office, 1974.

International Energy Agency. *World Energy Outlook*. Paris: OECD/IEA, 1982.

Jain, Kamal C., and R. J. P. de Figueiredo, eds. *Concepts and Techniques in Oil and Gas Exploration*. Tulsa: Society of Exploration Geophysicists, 1982.

King, Robert E., ed. *Stratigraphic Oil and Gas Fields—Classification, Exploration Methods, and Case Histories*. Tulsa: American Association of Petroleum Geologists, 1972.

Knowles, Ruth Sheldon. *First Pictorial History of the American Oil and Gas Industry, 1859-1983*. Athens, Ohio: Ohio University Press, 1983.

Landes, Kenneth K. *Petroleum Geology*. 2d ed. Huntington, N. Y.: Krieger Publishing, 1959.

Larson, Edwin E., and Peter W. Birkeland. *Putnam's Geology*. 4th ed. New York: Oxford University Press, 1982.

Leeder, M. R. *Sedimentology*. Boston: George Allen & Unwin, 1982.

Leet, L. Don, S. Judson, and M. E. Kauffman. *Physical Geology.* 6th ed. Englewood Cliffs, N. J.: Prentice-Hall, 1982.

LeRoy, L. W., and J. W. Low. *Graphic Problems in Petroleum Geology.* New York: Harper & Brothers, 1954.

Levorsen, A. I. *Geology of Petroleum.* 2d ed. San Francisco: W. H. Freeman and Company, 1967.

Link, Peter K. *Basic Petroleum Geology.* Tulsa: Oil and Gas Consultants International, 1982.

Markovskii, N. I. *Paleogeographic Principles of Oil and Gas Prospecting.* New York: John Wiley and Sons, 1978.

Mason, Brian. *Principles of Geochemistry.* New York: John Wiley and Sons, 1966.

McClean, J. G., and W. B. Davis. *Guide to National Petroleum Council Report.* Washington: National Petroleum Council, 1972.

McPhater, Donald, and Brian MacTeirnan. *Well-Site Geologists Handbook.* Tulsa: PennWell Publishing, 1983.

McQuillin, et al. *An Introduction to Seismic Interpretation.* Houston: Gulf Publishing, 1979.

Megill, Robert E. *An Introduction to Exploration Economics.* 2d ed. Tulsa: PennWell Books, 1979.

Merkel, Richard H. *Well Log Formation Evaluation.* Tulsa: American Association of Petroleum Geologists, 1979.

Mintz, Leigh W. *Historical Geology.* Columbus: Charles E. Merrill Publishing, 1981.

Moody, Graham B., ed. *Petroleum Exploration Handbook.* New York: McGraw-Hill, 1961.

Moore, Carl A. *Handbook of Subsurface Geology.* New York: Harper & Row, 1963.

Moses, V., and D. G. Springham. *Bacteria and the Enhancement of Oil Recovery.* London: Applied Science Publishers, 1982.

Myers, Donald A., P. T. Stafford, and R. J. Burnside. *Geology of the Late Paleozoic Horseshoe Atoll in West Texas.* Austin: The University of Texas, 1956.

National Petroleum Council. *Impact of New Technology on the U. S. Petroleum Industry, 1946–1965.* Washington: National Petroleum Council, 1967.

——. *U. S. Energy Outlook.* Washington: National Petroleum Council, 1972.

Nehring, Richard. *The Discovery of Significant Oil and Gas Fields in the United States.* Santa Monica, Cal.: Rand Corporation, 1981.

Nelson, H. Roice, Jr. *New Technologies in Exploration Geophysics.* Houston: Gulf Publishing, 1983.

Newell, Norman D., et al. *The Permian Reef Complex of the Guadalupe Mountains Region, Texas and New Mexico.* San Francisco: W. H. Freeman and Company, 1953.

Newendorp, Paul D. *Decision Analysis for Petroleum Exploration.* Tulsa: PennWell Books, 1975.

Nowacki, P., ed. *Oil Shale Technical Data Handbook.* Park Ridge, N. J.: Noyes Data Corporation, 1981.

Owen, Edgar W. *Trek of the Oil Finders: A History of Exploration for Petroleum.* AAPG Memoir #6. Tulsa: American Association of Petroleum Geologists, 1975.

Petroleum Information Corporation. *The Williston Basin—1980.* 1980.

——. *Overthrust Belt.* 1981.

Pirson, Sylvain J. *Geologic Well Log Analysis.* Houston: Gulf Publishing, 1977.

Pope, Clarence. *An Oil Scout in the Permian Basin.* El Paso: Permian Press, 1972.

Pratt, Wallace E., and Dorothy Good, eds. *World Geography of Petroleum.* London: Oxford University Press, 1950.

Priess, William G. *Understanding the Oil Business.* Scottsdale, Ariz.: Azoco, 1981.

Reineck, H. E., and I. B. Singh. *Depositional Sedimentary Environments.* 2d ed. New York: Springer-Verlag, 1980.

Rider, Don K. *Energy: Hydrocarbon Fuels and Chemical Resources.* John Wiley and Sons, 1981.

Rintoul, William. *Spudding In: Recollections of Pioneer Days in the California Oil Fields.* San Francisco: California Historical Society, 1976.

Roberts, John L. *Introduction to Geological Maps and Structures.* Oxford: Pergamon Press, 1982.

Robinson, Edwin S. *Basic Physical Geology.* New York: John Wiley and Sons, 1982.

Rocks, L. *Fuels for Tomorrow.* Tulsa: PennWell Publishing, 1980.

Rudd, Robert D. *Remote Sensing: A Better View.* Belmont, Cal.: Duxbury Press, 1974.

Ruedisili, Lon C., and Morris W. Firebaugh. *Perspectives on Energy.* 3d ed. New York: Oxford University Press, 1982.

Sawkins, F. J., et al. *The Evolving Earth.* 2d ed. New York: Macmillan, 1978.

Scholle, Peter A. *A Color Illustrated Guide to Carbonate Rock Constituents, Textures, Cements and Porosities.* Tulsa: American Association of Petroleum Geologists, 1978.

Selley, Richard C. *An Introduction to Sedimentology.* London: Academic Press, 1976.

Shelton, John S. *Geology Illustrated.* San Francisco: W. H. Freeman and Company, 1966.

Silver, Burr A. *Techniques of Using Geologic Data.* 2d ed. Tulsa: Institute for Energy Development, 1979.

——. *Subsurface Exploration Stratigraphy.* Oklahoma City: Institute for Energy Development, 1983.

Smith, William L., ed. *Remote-Sensing Applications for Mineral Exploration.* Strondsburg, Penn.: Dowden, Hutchinson and Ross, 1977.

Stokes, W. Lee. *Essentials of Earth History.* Englewood Cliffs, N. J.: Prentice-Hall, 1982.

Thomas, G. W. *Principles of Hydrocarbon Reservoir Simulation.* Boston: International Human Resources Development Corporation, 1982.

Tiratsoo, E. N. *Oilfields of the World.* 2d ed. Houston: Gulf Publishing, 1976.

Townshend, J. R. G., ed. *Terrain Analysis and Remote Sensing.* London: George Allen & Unwin, 1981.

Uren, L. C. *Petroleum Production Engineering.* 2d ed. New York: McGraw-Hill, 1934.

Vance, Harold. *Elements of Petroleum Subsurface Engineering.* St. Louis: Educational Publishers, 1950.

Waples, Douglas. *Organic Geochemistry for Exploration Geologists.* Minneapolis: Burgess Publishing, 1981.

Weston, J. Fred, and Eugene F. Brigham. *Managerial Finance.* 5th ed. Hinsdale, Ill.: Dryden Press, 1975.

Wrather, W. E., and F. H. Lahee. *Problems of Petroleum Geology.* Tulsa: American Association of Petroleum Geologists, 1934.

Wright, W. Floyd. *Petroleum Geology of the Permian Basin.* West Texas Geological Society, 1979.

INDEX

GREYSCALE

BIN TRAVELER FORM

Cut By _____Ed_____ Qty _42_ Date _2/13/25_

Scanned By_____ Qty_____ Date_____

Scanned Batch IDs

_____   _____   _____

Notes / Exception

_____

_____

_____

_____

_____